Routledge Revivals

Analyzing Tabular Data

First published in 1993, *Analyzing Tabular Data* is an accessible text introducing a powerful range of analytical methods. Empirical social research almost invariably requires the presentation and analysis of tables, and this book is for those who have little prior knowledge of quantitative analysis or statistics, but who have a practical need to extract the most from their data. The book begins with an introduction to the process of data analysis and the basic structure of cross-tabulations. At the core of the methods described in the text is the loglinear model. This and the logistic model, are explained and their application to causal modelling, to event history analysis, and to social mobility research are described in detail. Each chapter concludes with sample programs to show how analysis on typical datasets can be carried out using either the popular computer packages, SPSS, or the statistical programme, GLIM. The book is packed with examples which apply the methods to social science research.

Sociologists, geographers, psychologists, economists, market researchers and those involved in survey research in the fields of planning, evaluation and policy will find the book to be a clear and thorough exposition of methods for the analysis of tabular data.

Analyzing Tabular Data

Loglinear and logistic model for social researchers

Nigel Gilbert

Routledge
Taylor & Francis Group

First published in 1993
by UCL Press

This edition first published in 2022 by Routledge
2 Park Square, Milton Park, Abingdon, Oxon, OX14 4RN
and by Routledge
605 Third Avenue, New York, NY 10017

Routledge is an imprint of the Taylor & Francis Group, an informa business

Publisher's Note
The publisher has gone to great lengths to ensure the quality of this reprint but points out that some imperfections in the original copies may be apparent.

Disclaimer
The publisher has made every effort to trace copyright holders and welcomes correspondence from those they have been unable to contact.

ISBN: 978-1-032-19536-0 (hbk)
ISBN: 978-1-003-25970-1 (ebk)
ISBN: 978-1-032-19540-7 (pbk)

Book DOI 10.4324/9781003259701

Analyzing tabular data
Loglinear and logistic models for social researchers

Nigel Gilbert
University of Surrey

UCL
PRESS

First published in 1993 by UCL Press

UCL Press Limited
University College London
Gower Street
London WC1E 6BT

ISBN:
1-85728-090-3 HB
1-85728-091-1 PB

British Library Cataloguing-in-Publication Data
A catalogue record for this book
is available from the British Library.

Typeset in 9/12 pt Palacio.
Printed and bound by
Biddles Ltd, King's Lynn and Guildford, England.

CONTENTS

PREFACE

This book is meant for social researchers who would like to know how to analyze data tables. It is for you if you do not think of yourself as a statistician and, like the majority of social researchers, are more interested in getting results than in the statistical methods themselves.

The book arose out of a sense of frustration that, although almost all social research that deals with quantitative data (including sociology, geography, social psychology, policy analysis and allied disciplines) involves the presentation of numbers in tables, most statistics courses spend most of their time discussing techniques such as regression, analysis of variance and factor analysis, which are of limited use for analyzing tabular data. Over the past twenty years, a growing body of work has been devoted to a family of methods specifically aimed at the analysis of tables, all based around the core notion of a loglinear model. This work has made a major impression on the study of social mobility (e.g. Hauser 1978, Goldthorpe 1980), but its relevance to other fields is not yet sufficiently appreciated.

Most books about this new family of models are aimed at those with considerable mathematical and statistical sophistication. This is not because the ideas themselves are especially complex or difficult to master. This book presents in a non-technical way the ideas needed to understand loglinear modelling and it explains how to carry out loglinear analyses. A little statistics is needed as background, but no more than what is covered by most first-year undergraduate courses in the social sciences (specifically, the ideas of sampling, the mean and standard deviation of a sample and a population, the normal distribution, statistical inference and standard error). It is also helpful to have some experience of handling survey data and a basic knowledge of a data management computer program such as the Statistical Package for the Social Sciences (SPSS). A good refresher on social statistics is Loether & McTavish (1992). Dale et al. (1988) and Gilbert (1992, Ch. 12) provide introductions to survey management.

Because it is the practical application of loglinear analysis to tables that is the focus of the book, there is little statistical theory. The theory can be found in the references mentioned at the end of each chapter under the heading, "Further reading". The chapters include many illustrative examples of loglinear models, and all have a section on analysis procedures using either of the two most popular programs for loglinear analysis: SPSS and the Generalized Linear Interactive Modelling program, GLIM. A disk containing machine-readable copies of all the programs in the book is available.[1]

For the teacher of social statistics, the book is the equivalent of a one-term, ten-lecture course at roughly one chapter a week, plus some "optional extras" on more complex models at the end of the book. The material has been taught to second-year undergraduates following a BSc degree course in sociology at the University of Surrey and to MSc students following a course in social research methods.

The book begins by discussing the idea of modelling as a method of understanding social scientific data. It is important to see data analysis not as merely a matter of turning wheels until the answer pops out, but as a creative and theoretically informed process. The second chapter considers the different types of quantitative data available to social scientists and suggests why constructing tables is such a common form of analysis in social science. The basic structure of a two-variable table is described, and this is extended to tables involving more than two variables in Chapter 3. In the fourth chapter, the ideas of association and interaction are introduced as ways of measuring the strength of relationships manifested in tables. This brings the discussion to loglinear analysis, a method of modelling relationships between two or more variables, which is described in Chapter 5. There are often many models that can be fitted to a data table and so techniques for choosing between candidate models are needed, the subject of Chapter 6. One influential application of loglinear models, social mobility, is introduced in Chapter 7, which also describes some of the models that have been found useful in understanding mobility.

One of the major benefits of loglinear analysis is that it can be applied as easily to tables involving four, five or even more variables as to tables cross-classifying just two variables. Examining several variables at once allows more sophisticated analysis without much further statistical complication. Chapter 8 discusses such tables, and Chapter 9 considers how measures of relationship can be estimated from them.

The standard loglinear models treats all variables equally, but it is often more important to explain differences on one variable, the dependent or response variable, in terms of several other variables, the independent or explanatory variables. A variation of the loglinear model, called the logistic regression model, is described in Chapter 10 and is used for studying causal relationships in Chapter 11.

The standard loglinear models are best suited to data where the categories used for classifying are not naturally ordered. Chapter 12 shows how models can be developed for data where the categories are ordered, as they often are when measuring, for example, attitudes or educational achievement. The final chapter considers the application of logistic models to time-related data such as the date of marriage or gaining a job.

This book is based on an earlier volume, *Modelling society* (1981). The change in title for this major revision reflects a moderation of youthful enthusiasm. But the message of that title – that models are a vital ingredient for the development of sociological understanding – is also central to this edition. Since the earlier book was published, there have been substantial advances in loglinear analysis and related areas, and this edition covers some of the most important of these. Logistic regression now has a separate chapter, as does the extension of loglinear models to ordinal variables. A chapter on discrete-time event history analysis using logistic models has been added because of the increasing importance of event-based data with the development of longitudinal surveys. And some aspects of the new approach to mobility analysis based on topological models are briefly covered, although it is difficult to do justice to the vast literature that now exists on the topic.

In preparing the book, I have been encouraged and assisted by Bernadette Hayes at the University of Surrey, and by Colin Mills, now at the London School of Economics, both of whom kept me well supplied with books and articles. As ever, I benefited from discussions with Sara Arber and my other colleagues in the Department of Sociology at Surrey. I am particularly grateful to Colin Mills, Robert Erikson and Sara Arber for the data on which some of the examples of analyses are based. Sara Arber, Jane Fielding, Jay Ginn, Bernie Hayes, Clive Payne and Mike Procter read and commented on the manuscript and gave me good advice.

In *Modelling society*, I also expressed a debt of gratitude to a computer, a Prime "mainframe", for its help in computation and for its text-processing system. Fourteen years later, I am glad to acknowledge another computer, an Apple Macintosh, and to thank its designers for helping to make the process of writing of this book so much easier than the first version.

Note

1 The IBM PC compatible 3.5 inch disk is available for £10 including postage and packing from Nigel Gilbert, Department of Sociology, University of Surrey, Guildford, Surrey, GU2 5XH, England.

In memory of
Cathie Marsh

CHAPTER ONE

Real and imaginary worlds

Table 1.1 shows some data from a survey of a community in Oldham. People were asked whether they owned or rented their homes, and whether their parents had owned or rented theirs. What does the table show? The numbers in the rows and columns concerned with owner-occupiers are on the whole larger than the rest. There are several zeros, and these fall in particular areas of the table. These are just two instances of the patterns that can be found from a quick visual inspection. There may be other, more complex, patterns in these data which are not so easy to see and which need more sophisticated methods, involving the use of statistical techniques, to reveal.

Researchers are interested in patterns of numbers in data because

Table 1.1 Intergenerational change in housing tenure.

	Respondent's present form of tenure						
	Owner-occupied	Rented from			Other	Don't know, question not asked	Total
Parents' tenure		Local council	Private landlord, unfurnished	Private landlord, furnished			
Owner-occupied	84	2	7	12	0	0	105
Rented from council	39	3	2	1	1	1	47
Unfurnished, privately rented	77	5	13	0	3	0	98
Furnished, privately rented	4	0	2	0	0	0	6
Other	18	1	1	0	0	0	20
Don't know, not asked	117	2	26	4	1	0	150
Total	339	13	51	17	5	1	426

Source: derived from Crosby (1978: Table 21).

they provide clues to underlying social processes. For instance, the present housing tenure of the Oldham respondents might be influenced to some degree by whether their parents owned or rented their houses. This seems a likely possibility, given the fact that many other aspects of people's status and life-style are inherited. If there is indeed a tendency for parents' tenure to influence sons' and daughters' tenure, this ought to be reflected in the pattern of numbers in Table 1.1. Specifically, the values along the diagonal stretching from the top left to the bottom right ought to be larger than one would otherwise expect. In later chapters, we shall learn how to calculate whether this is the case, and so discover whether there is intergenerational inheritance of tenure.

One approach to examining tables is to try to find all the patterns of numbers in them. There is something to be said for this as a starting-point, especially if you know little about the social processes that might have generated the data. However, there are usually many patterns to be found in *any* data table. Some of these will be true reflections of interesting social processes. Others will be spurious, appearing quite by chance and reflecting nothing in particular.

This approach is like looking for patterns among the stars in the night sky and finding the signs of the zodiac – stars that appear to be outlining a ram, a crab, a lion and so on. We know that in fact the zodiac pattern is spurious, reflecting no real structures in the universe. In much the same way, we can always find some pattern in data, but finding it does not establish that the pattern is of any theoretical or substantive significance.

A better way of examining data is to draw on prior theoretical knowledge and insight about the social processes that might be involved, and to test whether the data show the consequences of those processes. In other words, to guide the search for patterns in the data we can use a theoretical *model* of what might be happening. A model is a theory or set of hypotheses which attempts to explain the connections and interrelationships between social phenomena. Models are made up of *concepts* and *relationships* between concepts (Gilbert 1992, Ch. 2).

The idea of a model can be illustrated with an example involving a very simple relationship. Suppose that you are studying the effect of education on social status. Education and social status are the two concepts. Theoretical ideas would suggest that, other things being equal, the higher one's educational attainment, the higher one's status.

The model is, then, that "status and education are directly related". It would be reasonable to have confidence in the validity of this model if data on education and status were obtained and it was found that, on the whole, those with more education had higher status.

In practice, however, education and status cannot easily be measured. Instead, *indicators* of the concepts would be examined. Possibly adequate but rather crude indicators in this instance might be the number of years at school or other educational establishments (to measure education), and the rating of current occupation on a scale of occupational prestige (to measure status). If we collected data using these indicators and found that the data showed a direct relationship between the two, this could be taken as evidence that the model is correct. We would have found a *structural correspondence* between the relationship proposed in the model and the relationship discovered in the data. Figure 1.1 illustrates the connections between social processes, a model and the data obtained from indicators.

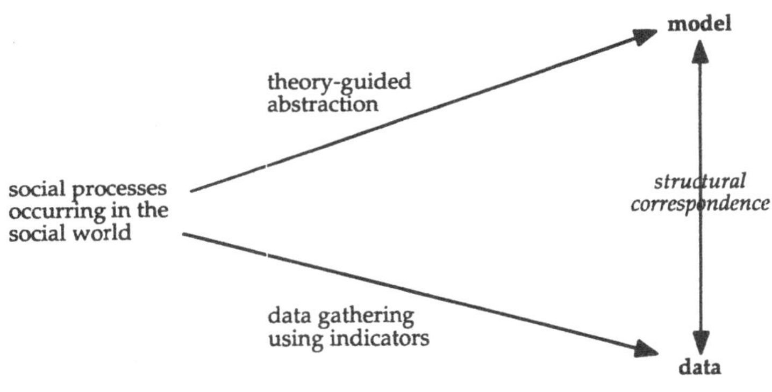

Figure 1.1 The relationship between a model and data.

Let's suppose that we have found such a structural correspondence. What does this mean? It suggests that the relationship specified in the model may exist in the "real world" or, to put it another way, that the model can represent that portion of the "real world" we have studied. The model can be used to make predictions about how the "real world" will respond to changes. Moreover, the relationships specified in the model will also serve as an explanation of how the "real world" works.

Nevertheless, it is necessary to be rather cautious about jumping to the conclusion that a model is a correct representation of the world from the discovery of structural correspondence alone. Such correspondence only provides evidence in support of the model, not definite confirmation of its validity. This is because the chain of argument leading to a demonstration of correspondence is a long and fairly complicated one, involving the derivation of a model from theory, the choice of indicators appropriate for testing the model, the collection of data using those indicators and, finally, a decision on whether correspondence has or has not been found. There is room for error in each of these steps.

Moreover, even if a model is correct, it never provides the full story. Inevitably, all models are simplified representations of the "real world". Not every pertinent relationship can be included in the model. For instance, educational attainment is certainly not the only determinant of status, although it may be the most important. A wide range of other factors influences status. A model can provide only a partial explanation and can give only an imperfect guide to prediction. This is of course especially true of a model as simple as the education and status example above, but it is also the case with the considerably more complicated models we shall meet in later chapters.

With these general ideas in mind, let us now look more closely at how one establishes whether there is structural correspondence between models and data. Models can be thought of as being located in an "imaginary world". This is identical in all respects to the "real world", except that the "imaginary world" includes the relationships specified in the model. Thus, the "imaginary world" is the world that would exist if the model were true. For instance, to return to the earlier example, in order to test the education - status model, you create in imagination a world identical to your own, except that in this world education *is* directly related to status.

Now you can compare the "imaginary world" with the "real world". If the two are indistinguishable, this is evidence in favour of the model. If the "imaginary world" differs from the "real world" (in which the relationships are the real but unknown ones), this is evidence that the model is incorrect. The problem of establishing structural correspondence is, therefore, reduced to the problem of comparing the "real" and the "imaginary" worlds. The comparison is performed by making measurements in both worlds. Data from the "real world" are obtained by observation, questionnaires and other collection procedures. Data from the "imaginary world" are obtained using data analysis techniques such as regression, factor analysis, loglinear analysis or whatever is appropriate. These techniques generate the data (often called *expected* or *fitted* data) that would have been collected if the "imaginary world" had really existed.

Thus we obtain two datasets, one from the real observation of the "real world" and one from the analytic technique that has been used to simulate the collection of data from the "imaginary world". If the two sets of data are identical, or are at least sufficiently nearly identical, this provides evidence for supposing that the "real" and "imaginary" worlds are in fact the same; that is, the model may correctly represent the true state of affairs. The connection between these worlds is illustrated in Figure 1.2.

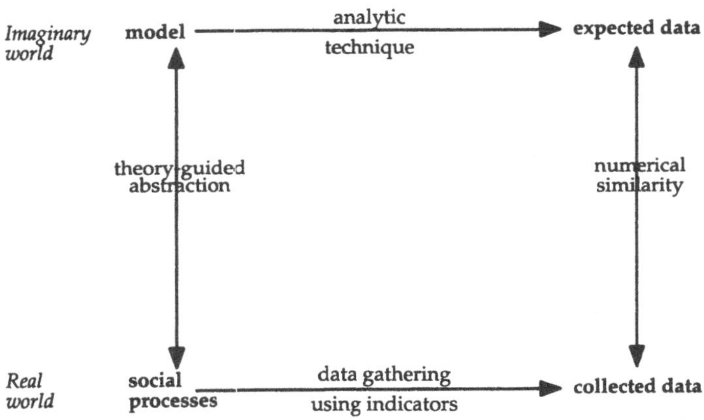

Figure 1.2 Schematic diagram illustrating the relationship between the "real" and "imaginary" worlds.

At this point, we can set out the steps needed to analyze a set of data:

1. From prior theoretical knowledge and the investigator's own insights, a model is designed which includes the important concepts and their interrelationships.
2. A choice is made about the analytical technique to be used. This will be determined by the form of the model that can be obtained and the sampling design used to select respondents.
3. The model is restated in terms appropriate to the analytic technique used.
4. The model and analytic technique are used in combination to generate a set of "expected" data (the data from the "imaginary world").
5. The "expected" data are compared with the data actually collected.

6 A decision is made about whether the model is acceptable, depending on whether the expected and observed datasets are sufficiently similar.
7 If the two datasets are not sufficiently similar, the model is rejected and steps 1–7 are repeated with another model, designed taking note of the differences found between the data sets.
8 If the model does appear to be acceptable, it is examined to see whether it can be simplified and refined. A simpler model is always preferable to an unnecessarily complex one. The process of refinement will involve testing the acceptability of simpler models, following steps 1–7.
9 By this stage it is known that certain relationships exist, but it is also useful to discover the relative importance or strength of these relationships. This is done by calculating the parameters of the model.

To validate the model, it is necessary to test it on another independent set of data. However, a second set of data is not often available and so it is rarely practical for the analyst to perform a validation. More usually, other researchers play the major role in validating models. Sometimes this may take the form of a strict replication of the analysis on other data. Much more frequently, the model that has been developed to explain data taken from one context is tested on data derived from different but related contexts.

This scheme of analysis differs in two main ways from the classical approach. First, statistical texts, especially ones pitched at an elementary level, generally consider the testing of hypotheses rather than models. The term "model" has been used here partly because it seems clearer than "hypothesis", and partly because, while a hypothesis usually concerns just one relationship, a model may and usually does involve a complex set of linked relationships. Secondly, the classical approach assumes that the analyst possesses, *a priori*, a carefully formulated hypothesis to be tested against the data. Following the confirmation or rejection of this hypothesis, the analyst must cease working with the original set of data. Improved hypotheses should be tested with new data. In contrast, the above scheme assumes that, while the analyst should have some prior theoretical notions about the form of suitable models, the investigation ceases only when an adequate model to describe the one set of data has been found. The task is essentially to explore in depth the structure of the data. This seems a much more realistic view of research in practice than the classical one.

Summary

Data analysis seeks to construct and test models against data collected from the social world. The models are intended to account for theoretically significant patterns in the data. In order to assess the validity of a model, an analytic technique is used to generate expected data, i.e. the data that would have been obtained if the model did correctly represent the real world, and these data and the observed data are compared. Successive models are then examined to find the model that fits best.

Using computer programs for analysis

Data analysis is almost always carried out with the help of a computer program because trying to calculate even the simplest models by hand is impossibly tedious and prone to error. There are a number of good programs available on the market, differing mainly in their ease of use and the range of analytic models they support. Among social researchers, probably the most widely used is SPSS (the Statistical Package for the Social Sciences).[1] This can be obtained in versions for mainframe computers and for personal computers (IBM compatibles and Apple Macintosh). The program is sold as a basic package to which additional modules can be added at extra cost. In order to carry out the analyses described later in this book, you will need the "Advanced Statistics" module.

The GLIM program (Generalized Linear Interactive Modelling) is a much more specialized program. It is the statisticians' favourite and is capable of analyzing all the models mentioned in this book and many others. It will run on either mainframe or personal computers.[2]

At the end of each chapter, there is a section describing how the methods outlined in that chapter can be put into practice with these two programs. Examples of the commands that should be used to do analyses, brief explanations of what the commands mean and sample output from the programs are provided. However, there is not space to explain all the details of how to operate the two programs; basic introductions for SPSS can be found in Norusis (1990b) and Frude (1987), and for GLIM in Healy (1988). The examples in this book are for version 4 of SPSS/PC+ and version 3.77 of GLIM. GLIM, version 4, has recently been released (Frances et al. 1993). This includes some new features, but all the GLIM examples in this book will run unchanged.

Comparing observed and expected data

Table 1.1 showed the actual patterns of intergenerational inheritance of tenure in Oldham (data from the "real world"). Chapter 4 will describe how you can calculate the table that would be expected in an "imaginary world" in which there was no inheritance of tenure – that is, a world in which people's own tenure was completely unaffected by their parents' tenure. For the moment, we shall have the computer programs calculate this table for us. The numbers in Table 1.1 can then be compared with those produced by the programs to see if they differ, and by how much.

SPSS

The SPSS set of commands in Program 1.1 will display a table similar to Table 1.1, but with both the actual counts and the counts that would be expected if there was no relationship between parents' and respondents' tenure.

Program 1.1

```
DATA LIST FREE /ParTen ResTen Count.
BEGIN DATA.
1 1  84    1 2 2    1 3  7    1 4 12    1 5 0    1 6 0
2 1  39    2 2 3    2 3  2    2 4  1    2 5 1    2 6 1
3 1  77    3 2 5    3 3 13    3 4  0    3 5 3    3 6 0
4 1   4    4 2 0    4 3  2    4 4  0    4 5 0    4 6 0
5 1  18    5 2 1    5 3  1    5 4  0    5 5 0    5 6 0
6 1 117    6 2 2    6 3 26    6 4  4    6 5 1    6 6 0
END DATA.
VARIABLE LABELS
 ParTen 'Parents Tenure' ResTen 'Respondents Tenure'.
VALUE LABELS ParTen ResTen
 1 'Owner' 2 'Council' 3 'Priv Unfurnished'
 4 'Priv Furnished' 5 'Other' 6 'DK, Not asked'.
MISSING VALUES ParTen ResTen (6).
WEIGHT BY Count.
CROSSTABS TABLES= ParTen BY ResTen
 /CELLS= COUNT EXPECTED.
```

The DATA LIST command names three variables: ParTen, the parents' tenure, ResTen, the respondents' tenure, and a variable, Count, to hold the counts in the body of the table. The data follow. Working cell by cell through the table, the rows of data give the category, numbered from 1 to 6, of the parents' tenure, the category number of the respondents' tenure, and then the count of the respondents with that particular combination of categories.

The next commands, VARIABLE LABELS and VALUE LABELS, provide descriptive labels for the variables and the rows and columns of the table. The MISSING VALUES command declares that for both variables the last category (Don't know (DK), Not asked) is to be ignored in constructing the table. The WEIGHT command specifies the relative weight to be given to each cell in the table; in this case, it is just the number of respondents falling into each cell. Finally, the CROSSTABS command requests SPSS to calculate and display the number of respondents in each cell (that is, as in the original table) and the number of respondents in each cell that would be expected if there were no relationship between parents' and respondents' tenure. The output from SPSS will include a table (see Table 1.2) in which each cell contains two numbers, the upper one being the original count, as in Table 1.1, and the lower one the expected count if the respondents' tenure were uninfluenced by parents' tenure. The numbers in each cell can be compared to see whether the "real" and "imaginary" worlds differ.

Table 1.2 Counts and expected values for the table of intergenerational change in housing tenure as displayed by SPSS.

PARTEN Parents Tenure by RESTEN Respondents Tenure

	Count Exp Val	RESTEN Owner 1.00	Council 2.00	Private Unfur- nished 3.00	Fur- nished 4.00	Other 5.00	Row total
Owner	1.0	84 84.8	2 4.2	7 9.5	12 5.0	0 1.5	105 38.2%
Council	2.0	39 37.1	3 1.8	2 4.2	1 2.2	1 .7	46 16.7%
Priv unfurnished	3.0	77 79.1	5 3.9	13 8.9	0 4.6	3 1.4	98 35.6%
Priv furnished	4.0	4 4.8	0 .2	2 .5	0 .3	0 .1	6 2.2%
Other	5.0	18 16.1	1 .8	1 1.8	0 .9	0 .3	20 7.3%
	Column Total	222 80.7%	11 4.0%	25 9.1%	13 4.7%	4 1.5%	275 100.0%

Number of missing observations: 151

GLIM

The commands given in Program 1.2 will perform a similar analysis using GLIM. The $units directive states that there are 25 cases to consider (the 25 cells in the body of Table 1.1, excluding the "Don't know, Not asked" categories and the row and column totals). The $factor directive names the row and column variables, or "factors", and states that both have five levels; that is, there are five different forms of tenure for both parents and respondents. $data indicates that the data will consist of values for the two factors and a variate, Count, which will hold the cell values. $read instructs GLIM to read the following data. Payne (1986) is the reference manual if you need further details on what these directives do, or, for version 4, Frances et al. (1993).

Program 1.2

```
$units 25
$factor ParTen 5 ResTen 5
$data ParTen ResTen Count
$read
1 1 84 1 2 2 1 3 7 1 4 12 1 5 0
2 1 39 2 2 3 2 3 2 2 4 1 2 5 1
3 1 77 3 2 5 3 3 13 3 4 0 3 5 3
4 1 4 4 2 0 4 3 2 4 4 0 4 5 0
5 1 18 5 2 1 5 3 1 5 4 0 5 5 0
$yvar Count
$error Poisson
$fit ParTen, ResTen
$display r
```

These data consist of three numbers for each cell: the first two specify the codes for the parent's and respondent's form of tenure, and the third, the count in that cell. The next three directives ($yvar, $error and $fit) specify the model to use. Chapter 4 will explain what these directives do. Finally, the $display directive asks GLIM to show the values expected in an "imaginary world" in which the model is true. The output (Table 1.3) consists of one line per cell of the table, giving the observed count, the "fitted" count (which is what GLIM calls the expected value) and the "standardised residual", a measure of the size of the difference between the observed count and the expected value (see Ch. 7).

Table 1.3 Counts, expected values and residuals for the table of intergenerational change in housing tenure as displayed by GLIM.

Unit	Observed	Fitted	Residual
1	84	84.764	−0.083
2	2	4.200	−1.073
3	7	9.545	−0.824
4	12	4.964	3.158
5	0	1.527	−1.236
6	39	37.135	0.306
7	3	1.840	0.855
8	2	4.182	−1.067
9	1	2.175	−0.797
10	1	0.669	0.405
11	77	79.113	−0.238
12	5	3.920	0.545
13	13	8.909	1.371
14	0	4.633	−2.152
15	3	1.425	1.319
16	4	4.844	−0.383
17	0	0.240	−0.490
18	2	0.545	1.969
19	0	0.284	−0.533
20	0	0.087	−0.295
21	18	16.145	0.462
22	1	0.800	0.224
23	1	1.818	−0.607
24	0	0.945	−0.972
25	0	0.291	−0.539

Further reading

The relationship between theory and data is one that has exercised sociologists and philosophers of science for many years. The view outlined in this chapter is developed more extensively in Willer (1967); Blalock (1982) considers the gap between theory and data from another angle. Abell (1971) advocates a "model-building perspective" and discusses the components of such models – concepts, variables, propositions and relations – in a careful but readable way. Gilbert (1992) relates data analysis to the social research process as a whole.

Note

1 Available from SPSS Inc., 444 N. Michigan Ave., Chicago, IL, 60611, USA, and local distributors.

2 GLIM can be obtained from Numerical Algorithms Group Ltd, Mayfield House, 256 Banbury Road, Oxford, OX2 7DE, England.

CHAPTER TWO
Classification and measurement

Quantitative data may be collected in many different ways, but common to all is a process of classification. For instance, there may be a need to count the number of events, people, or actions occurring in a particular social setting. But before counting, what is to be counted must first be classified: for example, to find out how many children there are in a school, those in the school must be divided into Children and Others (teachers, secretaries, cleaners, and so on).

Although classification may seem an uncomplicated task, a great deal depends on just how it is done. If the right categories are not chosen at the start, no amount of later analysis will come to the rescue. There are four main aspects of classification that need attention: the categories must all relate to some common property; the items to be classified into a category must be sufficiently similar with respect to that common property for them to be considered identical for analytical purposes; the categories must be mutually exclusive; and the set of categories must be exhaustive. Let us consider each of these in turn.

The categories chosen for a classification must all be concerned with the same idea. It is not sensible to ask "Are you male or aged over 21?" as an item in a survey. The question and the categories it suggests are based on two quite different variables: gender and age. Instead, two questions are needed – one about gender and one about age – and two separate classifications. A classification should be based on one, and only one, *property* of the items being classified.

It is not always as obvious as in this example that a set of categories relates to more than one property. Consider the categories listed in Table 2.1, which shows part of an early version of a question designed by the Office of Population Censuses and Surveys (OPCS) to measure ethnic affiliation. The list of "ethnic groups" seems to consist of a muddle of categories based on racial (for example, White), geographical (African), religious (Hindu) and cultural (Arab) characteristics (Mack

1978). Interpreted strictly, ethnic affiliation depends on culture, and the other properties - race, geography and religion - should have no part in a classification of ethnicity. The resulting potential confusion led OPCS to prepare improved versions of this question, intended for but not actually used in the 1981 Census.

Table 2.1 Question on ethnic affiliation.

10(1)	Race or ethnic group	
	Please tick the appropriate box to show the race or ethnic group to which the person belongs or from which the person is descended	1 ☐ White
		2 ☐ West Indian
		3 ☐ African
		4 ☐ Arab
		5 ☐ Chinese
		6 ☐ Indian
		7 ☐ Pakistani
		8 ☐ Bangladeshi
		9 ☐ Sri Lankan
		If Indian, Pakistani, Bangladeshi or Sri Lankan, please also tick one box below
		Hindi ☐ Sikh ☐ Muslim ☐ Other ☐

Source: adapted from Office of Population Censuses and Surveys (OPCS 1978: 6).

A second highly desirable feature of a classification is that each category should include only items that are the same or very similar with respect to the property being measured. For instance, until the 1971 Census, the Registrar-General classified occupations into five "classes", from I, which included "professional" occupations, to V, the "unskilled" occupations. But class III in the original scheme covered all skilled jobs, both manual and non-manual. The great diversity of occupations in class III meant that there was little that could be said in general about them. Recognizing this problem, since 1971 OPCS has subdivided class III into a non-manual and a manual component (OPCS 1970). Each of the new categories includes a more homogeneous group of occupations than the original large class.

Thirdly, the categories in a classification must be mutually exclusive; that is, all the items to be classified must find a place in only one of the categories. For example, when drafting a questionnaire it is easy to fall into the trap of categorizing Age into the ranges 16–25, 25–45, etc. This is a mistake, for those aged 25 can be placed in either the first or second category. The ranges must be altered to make them exclusive. Attitudinal questions also commonly raise difficulties. Table 2.2 shows data from a survey of working mothers with children aged under 11 who were asked to give the main reason why they had chosen to work. It is not clear that categories 2 and 6, and 3 and 5, are mutually exclusive. If the categories are not exclusive, the distribution of answers between them will depend not only on the respondents' attitudes, but also on the order in which the attitude statements occur on the questionnaire and on other irrelevant factors.

Finally, the set of categories must be exhaustive. Every item must be classifiable into one of the categories. In Table 2.2 the "Other" category was included solely to make the classification exhaustive; all but one of the respondents has chosen one of the reasons provided by the researchers. Similarly, a classification of family size might include "Ten or more children" as its last category. Because of their catch-all nature, very little is learnt about respondents who get classified into this kind of category, so classifications should be designed to reduce the "other" responses to a minimum.

Once the categories have been chosen and the classification scheme has been established, further decisions have to be made about the relationship between categories. As we shall see, the implications of these decisions are as important for the subsequent analysis as those made about the categories themselves.

Table 2.2 Main reason for mother with children choosing to go out to work.

(1) For the money	34
(2) To get me out of the house	33
(3) For the company	20
(4) To use my training	11
(5) Because I enjoy working	27
(6) Because I hate housework	3
(7) Other	1
Total	129

Source: survey for a student research project.

Non-metric and metric scales

The simplest type of classification is one in which the categories are just labelled, without any suggestion of a relationship between them. For instance, a classification by gender, in which people are categorized into males and females, is of this type, as is the ethnic groups classification in Table 2.1. Alternatively, the categories may be ordered, or they may be based on an underlying quantitative scale, as with the measurement of age or income. The relationship between categories is used to define the *level of measurement* of the scale.

The most basic level of measurement is the *categorical* (sometimes called the *nominal*) level. Measurement at the categorical level only involves applying names to the items being measured and, thus, classifying them. For instance, measurements might be made about people's political views. Respondents to a survey would be classified according to the appropriate party, "Labour", "Liberal Democrat", "Conservative" or "Other". The classification carries with it no implication of any ordering amongst the categories. Labour voters are not assumed to be better, bigger or longer than Conservatives, just different.

Measurement at the *ordinal* level means that the categories are ordered or ranked. As an example, respondents to surveys are frequently asked to indicate their opinion about an attitude statement by marking whether they strongly agree, agree, are indifferent, disagree, or strongly disagree. This is measuring their attitude at the ordinal level, for the classification is into categories that are assumed to have some intrinsic order. Another example of a question that yields an ordinal scale is shown in Table 2.3. It can be seen that the response categories are arranged in order, with a catch-all category, "Can't choose" at the end. Not all ordinal scales come from attitude questions such as this one, although they are perhaps the most common source. Social class is often measured at the ordinal level, as in the Registrar-General's classification of occupations mentioned above.

The feature of ordinal measurement that distinguishes it from the next higher level of measurement is that, although the categories are recognized to be ordered, no assumptions are made about the *amount* of difference between one category and the next. It is usually meaningless to ask about the amount of difference between categories on an ordinal scale. For instance, we do not assume that someone who responds "very satisfied" to a question is twice or any other quantifiable amount more satisfied than someone who replies that they are "fairly satisfied". Likewise, it does not make much sense to say that occupational class II has twice as much "classness" as class I.

Table 2.3 Question on university education.

				OFFICE USE ONLY
18.	Would you say that opportunities for university education are, in general, better or worse for women than for men?			
	PLEASE TICK ONE BOX	Much better for women	☐	1
		Better for women	☐	2
		No difference	☐	3
		Worse for women	☐	4
		Much worse for women	☐	5
		Can't choose	☐	9

Source: adapted from British Social Attitudes Survey, 1987 (Jowell et al. 1988).

Although we have been dealing with measurement, so far we have not used any numbers. What we have been doing is labelling categories with words such as 'very satisfied' or 'Hindu' to construct a scale, then assigning items to the categories according to some property of the items. This is sometimes described as *mapping* the property onto a scale. The relationship amongst the categories (as on an ordinal scale) should follow the relationship amongst the items being measured. If an ordinal scale is applied to items that cannot be compared, the mapping from items to scale becomes inaccurate: the scale indicates relationships that are not actually there.

It is equally desirable to choose a scale in which the relationships between the categories reflect the full relationships between the items. For example, if it is possible to compare and rank items on some property, a categorical scale should not be used because this loses the information available from the comparisons. The mapping in this case is not inaccurate, but it is incomplete.

There are many properties of items that can be not only compared but also added or subtracted. Common physical properties such as volume and mass are of this kind. Two volumes of beer, for instance, can be joined together ('added') to make a third, larger, volume. To measure such properties, a scale is needed in which the category labels can be both compared and added. The most convenient kind of scale with which to do comparison *and* addition is the one made up of numbers. In such a scale - a *metric* scale - the category labels are ordinary numbers.

The numbers represent the *amount* of the property possessed by the items being measured. For instance, we might measure the income of individuals on a metric scale, the numbers in this case representing the number of pounds sterling that make up a salary. Or we might measure the distance between respondents' homes and the nearest urban centre; here we would be likely to use a metric scale, in which the number assigned to a respondent would represent the number of miles out of town that he or she lives.

Metric scales are mappings of properties of items onto the number system. The mapping is appropriate only if the conventional relationships between numbers in the ordinary number system have their counterparts in the relationships between items. Two of the most fundamental operations that can be carried out with numbers are comparison and addition. Hence, a property ought to be mapped onto a metric scale only if it is meaningful to compare and to "add" (more formally, "concatenate") the items being measured. To apply a metric scale to a property, therefore, we need to be sure that items can meaningfully be both compared and concatenated. For instance, incomes can be compared and also concatenated, so we can use a metric scale to measure them.

The fact that, with a metric scale, a property has been mapped onto the number system conveys the great advantage that arithmetical procedures can be performed on the category labels. For instance, addition, subtraction and multiplication (though not always division) are meaningful operations on the labels - that is, the numbers of a metric scale. Since many powerful statistical techniques depend on such arithmetical procedures, it is best to measure on a metric scale whenever possible. However, addition is not meaningful for many sociologically interesting properties, and for these, non-metric scales are more appropriate.

Just as non-metric scales are divided into two types, categorical and ordinal, metric scales are also broken down into types, of which the most important is the *interval* level of measurement. An interval scale has all the features of a metric scale that we have discussed. Its categories are defined in terms of a base unit of measurement (the "interval"), such as pounds, miles or years of education. Items can then be classified into the categories according to the number of base units they possess.

A second type of metric scale is the *ratio* scale. This has the properties of an interval scale with, in addition, the feature that items exist that possess nothing of the property being measured. These items are classified into a "zero" category. For instance, households may be classified according to the number of children living in them. This is a ratio level measure, having a base unit or interval of one child, and possessing a zero category into which households with no children are placed. In contrast, social status is a concept that can in some circumstances be measured at the interval level, but that cannot be measured at the ratio level. (No one has "zero status".) Often, however, little depends on the difference between interval and ratio, and only the distinction between metric (including interval and ratio) and non-metric (including ordinal and categorical) measurement is significant.

A couple of important points of caution need to be made about these levels of measurement. Any category may be labelled with either a name or a number. Although applying numeric labels is often convenient, the numbers are simply a special kind of label for the categories. Only with metric measurements may arithmetical procedures on the numbers be used. Although the categories of "religious affiliation" could be labelled 1 for Protestant, 2 for Catholic and 3 for "other", it would not be sensible to calculate the average or mean religious affiliation of a number of people. The result might be 2.75, indicating somewhere between Catholic and other, a clearly nonsensical result. To calculate the mean, the category numeric labels have to be multiplied by the number of respondents in each category, an operation violating the logic of non-metric measurement. In contrast, it is legitimate to find the mean income of an occupational group, because in this case we are dealing with metric measurements for which the category labels relate directly to the number of intervals (pounds sterling) corresponding to those categories.

Another way of looking at this is that the numerical labels we applied to the categories of religious affiliation are arbitrary. We could alternatively have labelled them 101, 212 and 97, and these labels, although eccentric, would not be wrong. We do not have the same freedom with the income categories. These must be labelled with the correct quantities of pounds, because it is a metric scale. The categories of an ordinal scale may also be labelled with numbers, normally in a way that preserves the rank order of the categories; but again, ordinary arithmetical operations may not be used on these numbers. Only comparisons are valid.

Meaningless results are obtained when finding averages of non-metric measurements because the calculation involves using information not present in the data. Each level contains a differing amount of information, the least being in categorical-level measurements and the greatest in those at the ratio level. In general, there are considerable advantages in measuring at as high a level as possible, since then the most information is incorporated. However, sometimes measurements have to be made at an ordinal or categorical level because methods able to yield interval scales are not possible or have not been developed.

Often there are theoretically based reasons for regarding a measure as ordinal or categorical even though, at first sight, it appears to have been measured at the interval level. This is the case when a measurement is used as an indicator of a more fundamental, but difficult-to-measure concept. For instance, education is difficult to measure directly, if by 'education' is meant the quantity and quality of knowledge and skill people have acquired. We may be prepared to use the number of years that respondents have attended school and other educational establishments as an indicator of education, and "years of schooling" is obviously easy to measure at the interval level. But even so, the underlying concept, education, should be regarded as having been measured only at the ordinal level. Although the longer someone stays at school, the more they may be assumed to have learnt, the relationship between years of schooling and education cannot be made more precise than this. At present, few of the most important and fundamental concepts in social science can be measured at the interval level - even though some common indicators of those concepts can be.

Another example of measurement at the ordinal level, although the data would at first glance appear to be interval, occurs in the use of semantic differentials. Respondents are presented with a number of scales in the form of lines marked at each end with adjectives or *constructs* such as "friendly" and "unfriendly". They are asked to indicate the degree of friendliness of some person known to them by positioning a cross on the line. This is repeated for each construct and for a number of persons. We can easily measure the positions of the crosses on the lines to yield metric measurements. But it would be unwise to analyze such data as metric, for to do so would imply that respondents are able to construct a uniform scale between the "friendly" and "unfriendly" constructs and place their responses in terms of unit intervals along the line. A more reasonable procedure would involve dividing each scale line into coarse divisions and coding the positions of the crosses according to the divisions into which they fall. The information about the precise position of the crosses on the lines would be discarded as unreliable; only their relative ordering would be used in the analysis. In other words, the data would be treated as ordinal-level data.

The measurement process yields data on the number of "items" that is, people, events, actions or whatever – that fall into each category of a classification. This gives a *frequency* for each category. For example, Table 2.4 shows the frequency with which respondents chose the categories offered to them for the question about equal opportunities in university education shown in Table 2.3. These frequencies are simply counts measured at a ratio level. There is an important distinction between the level of measurement entailed in the classification of items into categories (which may be categorical, ordinal, interval or ratio) and the level of measurement of the counting once the classification has been performed (which is always at the ratio level). Because the counts are at the ratio level, we can calculate percentages, as shown in the table.

Table 2.4 Frequencies for the responses to a question on equal opportunities in university education.

	Count	%
Much better for women	14	1.1
Better for women	15	1.1
No difference	967	75.5
Worse for women	132	10.2
Much worse for women	10	0.8
Can't choose	143	11.1
Total	**1281**	**99.8**

Source: Jowell et al. (1988).

Levels of measurement have been discussed in some detail because it is very important not to use techniques of analysis that assume that one's data contain more information than they actually do. An example was given above of the worthless results obtained when a mean of the categorical variable, religious affiliation, is calculated. This was because, to compute a mean, information that is available only in metric data is required. Alternative procedures which do respect the lesser amount of information available in non-metric data have been developed to measure the 'central tendency' of a variable. The *mode* is used for categorical scales. The category containing the most items (the one in which the items occur with the greatest frequency) is the *modal category*. Table 2.5 shows data measured on a categorical scale for which the mode is the most appropriate way of measuring central tendency. Clearly, the modal category is that for "Married couple, no children". The *median*, the measure of central tendency for ordinal scales, is defined as that category which has half the items falling in categories within and below the median category, and half in those within and above. Table 2.6 shows data for which the median is the appropriate measure, if we assume that the categories of socio-economic status are ordered. The median category is "Skilled manual". Notice that the definition of the median requires that the categories can be compared and ordered (so that we can know which are above and below the median), but the definition of the mode does not require making comparisons or knowing anything about the interrelationships between categories. Hence, while the mode is available for any scale, a median can be obtained only for scales in which comparison is meaningful (ordinal or metric).

Table 2.5 Households by type, 1989.

	Number	%
Living alone	2541	25
Married couple, no children	2785	27
Married couple with dependent children	2614	26
Married couple with non-dependent children	884	9
Lone parent with dependent children	500	5
Other households	761	8

Source: Breeze et al. (1991, Table 2.23).

Table 2.6 Socio-economic profile of economically active local authority tenants, 1988–89.

	%
Professional and managerial	7
Intermediate and junior non-manual	12
Skilled manual, etc.	41
Semi-skilled manual, etc.	27
Unskilled	12

Source: derived from Breeze et al. (1991, Table 8.10.)

All statistical measures and procedures can be classified according to the demands they make on data. Some of the more common one- and two-variable procedures are listed in Table 2.7, grouped by the type of data they require. Those in the "metric" boxes are the best known and the most widely used.

This is not because most sociological data are metric. On the contrary, it is quite hard to find sociological concepts that satisfy all the requirements of metric measurement. Nevertheless, researchers often have been forced into using metric procedures for lack of more appropriate ones and have had to hope that they were not thereby generating entirely spurious results.

Table 2.7 Statistical techniques classified by the levels of measurement they require.

Level of measurement	Relationship between categories	Applicable measures of central tendency	Two-variable procedures Level of second (explanatory) variable		
			Categorical	Ordinal	Interval
Non-metric					
Categorical (nominal)	None	Mode	Loglinear analysis	Loglinear analysis	
Ordinal	Ordered	Median		Multi-dimensional scaling	
Metric					
Interval and ratio	Mapped onto real numbers	Mean	Analysis of variance		Regression Factor analysis

Regression, factor analysis and analysis of variance were first developed not in sociology, but in disciplines such as biology, agriculture and psychology, where metric data are more easily obtainable. They were imported into sociology because of their apparent utility. Although these techniques may be applied to some sociological research problems with valuable results, their usefulness is much more restricted than is commonly supposed because of their reliance on the extra information available in metric data.

The purpose of this book is to introduce a family of techniques that are particularly suitable for the analysis of sociological data, because they do not demand that measurements have been made using a metric scale. Loglinear analysis is most appropriately used with data measured at the categorical level, although it can also be used with ordinal data. But before we become involved in the details of these techniques, we must review some basic features of cross-tabulations, the starting-point for many non-metric analyses.

Summary

Measurement involves classifying items into categories and counting the number of items falling into each category. The categories should be chosen so that they relate to a common property and are mutually exclusive and exhaustive, and so that each category includes items that can be considered similar for analytic purposes. The relationship between categories is described in terms of levels of measurement: either categorical (no relationship); ordinal (categories are ordered); interval (categories are based on a number scale) or ratio (the categories are based on a number scale with a true zero). The first two levels of measurement are non-metric, the latter are metric. The level of measurement of a classification indicates the amount of information it incorporates, and this determines the arithmetical and statistical procedures that may be carried out on it. The most familiar statistical techniques require metric measurement, although most sociological data are non-metric.

Computer analysis

One of the great benefits of computer programs for data analysis is that they will quickly and accurately perform routine and tedious operations such as counting responses. They will also compute measures of central tendency and other statistics. However, it is left to the analyst to decide which of all the statistics that can be calculated are applicable to the data and useful for the problem at hand.

SPSS

SPSS will compute frequency counts with the FREQUENCIES command. It will also calculate measures of central tendency, regardless of whether they are meaningful. It is then the responsibility of the researcher to decide which measure is appropriate and how it should be interpreted.

The set of SPSS commands in Program 2.1 will read a list of respondents' ages (in years, one value per respondent) and display a frequency count, a histogram to show the distribution of ages visually, and the mode, median and mean of the data. Compare the three statistics against the histogram and see that they measure different aspects of "central tendency".

Program 2.1

```
DATA LIST FREE / Age.
BEGIN DATA.
53  20  60  61  88  27  25  65  49  42  22
19  76  64  64  60  38  40   3  69  36  34
10   8  45  37  30  25   2  35  15  13   4
20  18  30  28   2  56  56  38  33  11   9
 9   7   6  67  74  37  29   3   0  32  32
END DATA.
VARIABLE LABELS Age 'Age in years'.
FREQUENCIES VARIABLES = Age
  /HISTOGRAM
  /STATISTICS= MODE,MEDIAN,MEAN.
```

GLIM

GLIM does not provide a built-in procedure for calculating the mode. It will, however, compute the median and mean and display a frequency count and a histogram.

Program 2.2

```
$units 55
$data Age
$read
53  20  60  61  88  27  25  65  49  42  22
19  76  64  64  60  38  40   3  69  36  34
10   8  45  37  30  25   2  35  15  13   4
20  18  30  28   2  56  56  38  33  11   9
 9   7   6  67  74  37  29   3   0  32  32
$assign Cats = 0, 10, 20, 30, 40, 50, 60, 70, 80, 90, 100
$group GrpAge = Age intervals Cats
$tabulate for Age
$histogram GrpAge
$tabulate the Age fifty
$tabulate the Age mean
```

In Program 2.2, the $units directive indicates that there are 55 data items to be read into the Age variable, defined in the $data directive. The first step is to classify the data into ten-year age groups. The grouping is defined by the Cats variable, which is set up with the $assign directive. (Respondents aged between 0 and 9 years will be given the code 1; those aged 10–19 get a 2, and so on.) The $group directive computes a new variable, GrpAge, which holds the Age variable grouped into these ten-year categories.

The first $tabulate directive will show how many respondents there are in each year of Age category. The $histogram directive will create a chart showing the number of respondents in each age group. The second $tabulate will compute the median (the age at which 50 per cent of the sample are older) and the third, the mean.

Alternatively, you can use the "macro library" supplied with the GLIM package. A macro is a chunk of program that can be read into GLIM and then executed. The macro library is divided into a number of subfiles, with each subfile including a number of related macros. The SUMM subfile deals with summary statistics for a variable, including the macro SUMMARY, which prints the mean, variance, standard deviation, minimum and maximum, range, median and quartiles of a variable (but not the mode).

To use a macro from the standard library, you first have to read it into GLIM from the file in which it is stored. The following directive will achieve this (what the directive means does not matter much, but an explanation can be found in the GLIM manual):

```
$input %plc 80 SUMM $.
```

To use the macro to print summary statistics, enter the directive

```
$use SUMM Age$.
```

The complete program to read in and display statistics, using the library macro, is given in Program 2.3.

Program 2.3

```
$input %plc 80 SUMM $
$units 55 $
$data Age $
$read
53  20  60  61  88  27  25  65  49  42  22
19  76  64  64  60  38  40   3  69  36  34
10   8  45  37  30  25   2  35  15  13   4
20  18  30  28   2  56  56  38  33  11   9
9    7   6  67  74  37  29   3   0  32  32
$use SUMM Age$
```

Further reading

A brief, clear discussion of levels of measurement is to be found in Blalock (1979, Ch. 2). Abell (1971, Ch. 4) develops the idea of measurement systems in more depth. Oppenheim (1991) is good on question design and attitude scaling.

CHAPTER THREE
Cross-tabulations

Cross-tabulations, also called *contingency tables*, lie at the heart of most quantitative social research. Given a cross-tabulation such as Table 3.1, the aim is to discover whether there is a relationship between the variables and, if there is, to find its form. Once a relationship has been identified, it can be examined to see whether the data confirm or contradict a theory. Later, we shall be concerned with ways of modelling relationships between variables in cross-tabulations using a technique called loglinear analysis. First, however, some notation and terminology must be introduced.

Table 3.1 Mothers' preference for the gender of their next child.

	Present family		
Mother would prefer baby to be:	Boy(s) only	Girl(s) only	Boy(s) and girl(s)
Boy	24	210	30
Girl	209	24	21
Don't mind	105	100	47

Source: Cartwright (1976, derived from part of Table 16).

Table 3.1 shows a simple cross-tabulation of two variables: mothers' preferences for the gender of their next child, and the gender composition of the mothers' present families. There is a relationship between the two variables if the mothers' preferences vary according to the kind of family they have at present. It is clear that their preferences do, indeed, vary depending on their present family composition. On the whole, mothers with children of one gender tend to want their next child to be of the other gender, but those with both sons and daughters are more likely to have no strong preference for a boy or girl. As Cartwright (1976) notes, the relationship between family composition and preference for a son or daughter might have important demographic implications if it ever became possible for people to choose the gender of their children.

Table 3.1 is an example of a cross-tabulation of just two categorical variables. Let us now look at an example of a more complex table. The UK Marriage Research Centre conducted a survey of divorcees (Thornes & Collard 1979), and we shall use some unpublished data from this study as a continuing example of a multivariate cross-tabulation here and in Chapter 4. The kind of questions that can be answered using the data of Table 3.2, are: "Are those who say that they have had affairs while married also likely to admit to sex before marriage?"; "Are divorced men and women more likely to admit to having been involved in premarital sex or adultery than those still married?" and "Is there a difference in the willingness of men and women to report that they have been involved in sexual adventures?"

Table 3.2 Gender, by reports of experience of premarital sex (PMS) and extramarital sex (EMS), by marital status.

Gender	Reported PMS	Reported EMS			
		Divorced		Still married	
		Yes	No	Yes	No
Women	Yes	17	54	4	25
	No	36	214	4	322
Men	Yes	28	60	11	42
	No	17	68	4	130

Source: UK Marriage Research Centre.

The data in Table 3.2 were obtained by asking around 500 randomly selected men and women who had petitioned for divorce, and a similar number of married people two questions: (a) "Before you married your (former) husband/wife, had you ever made love with anyone else?" (b) "And during your (former) marriage, (did you have) have you had any affairs or brief sexual encounters with another man/woman?" Note that, although the researchers took some precautions to put their respondents at ease and assure them that their replies were in confidence, we should be careful not to place too much reliance on the accuracy of the answers as reports of fact. Divorcees might well have admitted to their sexual adventures more freely than those still married.

Since the analysis of cross-tabulations like Table 3.2 will be the focus of this chapter, it would be useful to begin by defining some terms to describe them. The table cross-classifies four *variables*, each of which is divided into *categories* or *levels* ("Yes" and "No", "Men" and "Women"). In this table all the variables have two categories (although in many cases variables are divided into more than two categories). Each category is identified by a *label*. Each combination of categories is represented by one *cell*. A cell contains a count of the number of respondents whose replies and whose gender and marital status correspond to that cell's combination of categories. Because the cells hold counts or frequencies, the table is said to be a *frequency table*. The number of variables in a table is often referred to as the table's *dimensionality*. Table 3.2 has been arranged in the square format most commonly found in the literature, although other arrangements can sometimes be clearer. Table 3.3, for example, is equivalent to Table 3.2 but has the variables rearranged so that marital status is laid out along one side with the other variables across the top.

Table 3.3 Marital status, by gender by reports of pre- and extramarital sex (PMS and EMS).

Marital status	Women				Men			
	Reported PMS				Reported PMS			
	Yes		No		Yes		No	
	Reported EMS		Reported EMS		Reported EMS		Reported EMS	
	Yes	No	Yes	No	Yes	No	Yes	No
Divorced	17	54	36	214	28	60	17	68
Still married	4	25	4	322	11	42	4	130

Marginal tables

If the frequencies along the top row of Table 3.3 are summed, the result gives the total number of divorcees in the sample as a whole. Similarly, the sum of the frequencies in the bottom row is equal to the number of married people in the sample. These totals are known as *marginals*, because they are conventionally added to the table along its margin. The total numbers of women and men can be found by summing the cell frequencies in the left- and right-hand halves of the table, respectively; and, by selecting the appropriate sets of cells, the totals of those responding one way or the other to the questions about pre- and extramarital sex can be calculated. These totals are also known as marginals, although there is no convenient margin in Table 3.3 on which to record them.

The marginals indicate the frequencies obtained by disregarding all the other variables. For instance, the marital status marginal obtained by adding along the rows represents the frequencies of married and divorced people in the sample, ignoring for the moment all the other information we possess about them. *Marginal tables* (sometimes also called just "marginals") can be constructed by disregarding all but a few of the variables. For instance, Table 3.4 is a marginal table cross-classifying marital status and gender, in which the information on respondents' reported experience of pre- and extramarital sex has been disregarded by "summing over" these latter two variables – that is, by adding each half-row of Table 3.3 to give the corresponding cell of the marginal table. Before reading on, you should satisfy yourself that you can obtain Table 3.4 from Table 3.3.

Table 3.4 Marginal table of marital status, by gender.

Marital status	Gender	
	Women	Men
Divorced	321	173
Married	355	187

Table 3.4 shows that there are roughly equal proportions of divorced men and women in the sample. Ten different marginal tables can be obtained by summing the data of Table 3.3 over various combinations of variables. Table 3.5 is another one of these, obtained by summing over the categories of the variable, marital status.

All the tables we have seen so far can be divided into subtables. Table 3.5, for instance, can be split into a table relating to men (the right-hand half) and a table relating to women (the left-hand half). Such subtables are known as *partial* or *conditional tables* – conditional in this instance on gender. Another way of saying much the same thing is to describe the two tables as showing the relationship between reported experiences of pre- and extramarital sex, *controlling for* the gender of the respondent. It is evidently rather easy to become overwhelmed by the great number of marginal and partial tables that can be extracted from multidimensional cross-tabulations. As we shall see, loglinear analysis is a useful technique for social researchers because it provides a powerful way of describing and investigating this multitude of tables.

Table 3.5 Marginal table showing reported experience of premarital sex (PMS), by experience of extramarital sex (EMS) and gender.

	Gender			
	Women Reported EMS		Men Reported EMS	
Reported PMS	Yes	No	Yes	No
Yes	21	79	39	102
No	40	536	21	198

Because in describing loglinear analysis we shall need to make general statements about cross-tabulations of any dimensionality and with any number of categories per variable, we need a way of referring to the cells in a *generalized table*. Using this convention, we shall be able to develop a uniform method for the analysis of any cross-tabulation. The generalized table has variables A, B, C, D and so on – an arbitrary number. We assume that variable A has a total of I categories labelled with subscripts, thus: A_1, A_2, A_3, A_4 and so on, to the last, A_I. The categories of variable B will be B_1, B_2, B_3 and so on, to the last category B_J. The remaining variables will have categories similarly labelled C_1, . . . , C_K, D_1, . . . , D_L, and so on. The capital letters I, J, K and L represent the number of categories of each of the variables A, B, C and D. We shall use the corresponding lower-case letters i, j, k and l as subscripts to represent any particular category of the appropriate variable. So the value of i can range between 1, meaning the first category of variable A, to I, the last category of variable A. Similarly, j, the subscript for variable B, can take any value between 1 and J.

The frequencies in the cells of the generalized table will be represented by the symbol x, with subscripts to indicate which particular cell frequency we are considering. Table 3.6 shows a four-dimensional example of a generalized table. This example has been constructed so that it looks like Table 3.3, but with the notation replacing the data. Often we shall need to refer generally to any cell of a table such as Table 3.6, rather than to one particular cell. In such cases, notation such as x_{ijkl} (that is, the frequency in the cell corresponding to category i of the variable A, category j of the variable B, category k of the variable C and category l of the variable D) means "any one of the cell frequencies in a four-dimensional table".

Table 3.6 Example of general notation applied to a four-dimensional table.

Marital status	Women Reported PMS				Men Reported PMS			
	Yes Reported EMS		No Reported EMS		Yes Reported EMS		No Reported EMS	
	Yes	No	Yes	No	Yes	No	Yes	No
Divorced	x_{1111}	x_{1211}	x_{1121}	x_{1221}	x_{1112}	x_{1212}	x_{1122}	x_{1222}
Still married	x_{2111}	x_{2211}	x_{2121}	x_{2221}	x_{2112}	x_{2212}	x_{2122}	x_{2222}

Recall that, to construct a marginal table, all the categories of one or more variables are summed over. For instance, to obtain the marginal table of marital status by gender shown as Table 3.4, we summed over all the categories of the variables concerned with experience of pre- and extramarital sex. To arrive at the cell frequency for the top left-hand corner (divorced/women), we added

$$17 + 54 + 36 + 214 = 321$$

To symbolize a variable that has been summed over, we use a plus sign (+) as a subscript instead of a lower-case letter. Hence, in generalized notation the above addition would become:

$$x_{1111} + x_{1211} + x_{1121} + x_{1221} = x_{1++1}$$

The plus signs in the result represent subscripts of variables that have been summed over. Note that it is only the subscripts of the second and third variables that change from term to term as we sum over these variables' categories. Next, let us look at the frequency in the top right-hand corner cell of the marginal table. In generalized notation, the calculation is:

$$x_{1112} + x_{1212} + x_{1122} + x_{1222} = x_{1++2}$$

(Make sure you follow which cells in the main table, Table 3.6, are being summed.) The marginal table as a whole is shown in Table 3.7, and this is the generalized equivalent of the marital status by gender table, Table 3.4.

Table 3.7 A marginal table in generalized notation.

	D	
A	D_1	D_2
A_1	x_{1++1}	x_{1++2}
A_2	x_{2++1}	x_{2++2}

To help make this notation for marginals completely clear, Table 3.8 reproduces Table 3.7 with the addition of all *its* marginals. See how the plus signs replace the subscripts being summed over. The table total is found by summing over *all* the variables (that is, adding together all the cell frequencies), and so its notation is x_{++++}.

Table 3.8 A marginal table with its marginals.

		D	Total
A	D_1	D_2	
A_1	x_{1++1}	x_{1++2}	x_{1+++}
A_2	x_{2++1}	x_{2++2}	x_{2+++}
Total	x_{+++1}	x_{+++2}	x_{++++}

All these definitions are not very exciting and perhaps a little confusing at first sight. However, they are essential if we are to develop methods of analysis of multivariate cross-tabulations that can be used on all tables, no matter how complex they may be. Using the notation, we shall be able to apply powerful methods which, once learnt, can be applied quite generally and in the same way to any cross-tabulation. Nevertheless, in Chapter 4 we begin in a more modest way by considering just the relationships between a pair of variables and use that as a basis for more complicated analyses.

Summary

A table cross-classifies two or more variables, each divided into several categories or levels. The cells of a table may hold counts (or frequencies) and possibly other measures such as percentages or expected values. Cells may be summed to give various marginals and marginal tables, that is those tables which would be obtained by disregarding one or more of the original tables' variables. There is a general notation, suitable for referring to any cell in a table with an arbitrary number of variables, which is important in developing methods for the analysis of complex cross-tabulations.

Computer analysis

Multivariate tables can be calculated and displayed by both SPSS and GLIM.

SPSS

Program 3.1 displays the data shown in Table 3.2. The DATA LIST command defines the four variables involved and an additional variable, Count, to hold the cell counts. This is followed by the rows of data, each consisting of a cell count and the levels of the four variables for

Program 3.1

```
DATA LIST FREE / MarStat, EMS, PMS, Gender, Count.
BEGIN DATA.
1 1 1 1   17
1 2 1 1   54
1 1 2 1   36
1 2 2 1  214
1 1 1 2   28
1 2 1 2   60
1 1 2 2   17
1 2 2 2   68
2 1 1 1    4
2 2 1 1   25
2 1 2 1    4
2 2 2 1  322
2 1 1 2   11
2 2 1 2   42
2 1 2 2    4
2 2 2 2  130
END DATA.
VARIABLE LABELS
   MarStat 'Marital Status'
   EMS 'Extramarital Sex'
   PMS 'Premarital Sex'.
VALUE LABELS
   MarStat 1 "Divorced" 2 "Married"/
   EMS, PMS 1 "Yes" 2 "No"/
   Gender 1 "Women" 2 "Men".
WEIGHT BY Count.
CROSSTABS TABLES = MarStat BY EMS BY PMS BY Gender
   /CELLS = COUNT, COLUMN.
```

that cell. The variables and their levels are then labelled. The last command requests a cross-tabulation including both the counts themselves and the counts percentaged down the columns of the table. By specifying other keywords, it is also possible to obtain percentages across rows and percentages over the whole table.

SPSS prints a separate two-dimensional partial table, cross-tabulating marital status (MarStat) and Extramarital sex (EMS) for each combination of values of the other two variables, Premarital Sex (PMS) and Gender. The first of these partial tables, that for women who reported premarital sex, is reproduced in Table 3.9. If the four partial tables were placed side by side, the arrangement of the cells would be rather like that shown in Table 3.3.

Table 3.9 Marital status, by extramarital sex for women reporting experience of premarital sex, from SPSS output.

```
MARSTAT Marital Status by EMS Extramarital Sex
Controlling for..
PMS Premarital Sex Value = 1.00 Yes
GENDER Value = 1.00 Women
```

	Count Col Pct	EMS Yes 1.00	 No 2.00	Page 1 of 1 Row total
MARSTAT				
Divorced	1.00	17 81.0	54 68.4	71 71.0
Married	2.00	4 19.0	25 31.6	29 29.0
	Column total	21 21.0	79 79.0	100 100.0

In the example in Table 3.9, the data took the form of cell counts. Often, it is necessary to tabulate "raw" data, that is, data in which each row represents one respondent (rather than one cell of a table). In this case, a row would specify the values taken for a respondent on each of the variables. To tabulate this kind of data, only small changes need to be made to the program, as shown in Program 3.2. The first data line (a data *record*) is 1 2 1 2 and represents a particular respondent who is divorced (MarStat is 1), did not report extramarital sex (EMS is 2), did report premarital sex (PMS is 1), and is female (Gender is 2). In all, there are 60 records in the data that have this pattern of responses and so the cell count for such respondents is 60 in Table 3.2. The differences between Program 3.1, which reads in cell counts, and Program 3.2, which reads in one record for each respondent, are that the former has a Count variable in the DATA LIST, cell counts in the data, and a WEIGHT command. The WEIGHT command has the effect of taking each pattern of responses and duplicating it Count times. Thus, as far as SPSS is concerned, it does not matter whether you read in "raw" data, record by record, or read cell counts to use as weights: the results of any subsequent analyses will be identical.

Program 3.2

```
DATA LIST FREE / MarStat, EMS, PMS, Gender.
BEGIN DATA.
1 2 1 2
1 1 1 2
2 1 2 1
2 2 1 1
2 2 1 2
1 2 2 1
.......
[One row for each respondent]
.......
2 1 1 2
1 1 1 1
2 1 2 1
1 2 1 1
2 2 1 2
END DATA.
VARIABLE LABELS
  MarStat 'Marital Status'
  EMS 'Extramarital Sex'
  PMS 'Premarital Sex'.
VALUE LABELS
  MarStat 1 'Divorced' 2 'Married'/
  EMS, PMS 1 'Yes' 2 'No'/
  Gender 1 'Women' 2 'Men'.
CROSSTABS TABLES = MarStat BY EMS BY PMS BY Gender
  /CELLS = COUNT, COLUMN.
```

GLIM

GLIM produces much more condensed tables than SPSS, but it is not able to label the variables or their categories with helpful legends, as SPSS does. In Program 3.3, which prints the data of Table 3.2, the $units directive specifies that there are 16 cells in the table. The $factor directive names the variables and indicates how many levels each has, and the $data directive identifies the data that follow the $read directive.

GLIM allows variables to have names with any reasonable number of letters, but it stores only the first four. For example, it treats the variable name, "Gender", as though only "GEND" had been typed. This means that to avoid errors all the variables used in a program must differ in their first four initial letters (eight in GLIM version 4).

The $tprint (Table print) directive specifies first the variable from which the counts in the table cells are to be taken (the variable Count) and then the variables that are to form the dimensions of the table, the latter separated one from the next by semi-colons. Three styles of output are available: −1, which just displays the counts without any identifying labels; 0, which adds the names of the cross-classifying variables and their levels; and 1, which adds graphic borders. Table 3.10 shows the output produced using style 1.

Program 3.3

```
$units 16
$factor MarStat 2 EMS 2 PMS 2 Gender 2
$data MarStat EMS PMS Gender Count
$read
1 1 1 1    17
1 2 1 1    54
1 1 2 1    36
1 2 2 1   214
1 1 1 2    28
1 2 1 2    60
1 1 2 2    17
1 2 2 2    68
2 1 1 1     4
2 2 1 1    25
2 1 2 1     4
2 2 2 1   322
2 1 1 2    11
2 2 1 2    42
2 1 2 2     4
2 2 2 2   130
$tprint (style = 1) Count MarStat; EMS; PMS; Gender$
```

Table 3.10 Output from GLIM: tabulation of data in Table 3.2.

	PMS	1		2	2
	GEND	1	2	1	
MARS	EMS				
1	1	17.000	28.000	36.000	17.000
	2	54.000	60.000	214.000	68.000
2	1	4.000	11.000	4.000	4.000
	2	25.000	42.000	322.000	130.000

Further reading

Zeisel (1958) is a good introductory guide to the construction and reading of cross-tabulations. Rosenberg (1968: App. A) provides a shorter treatment. Bryman & Cramer (1990) explain how to create and interpret cross-tabulations using SPSS.

CHAPTER FOUR
Association and interaction

We have already seen that a primary concern in analyzing a cross-tabulation is modelling the relationships between the variables. The relationship that might exist between two variables such as those shown in Table 4.1 is known as an *association*. In the first part of this chapter, we shall explore various ways of seeing whether there is association in a simple two-variable table. Then we consider the relationship known as *interaction* which may be found among three variables.

Table 4.1 Income, by occupation, for respondents who reported the income they earned in 1976.

Income	Occupation		
	Manual	Non-manual	Total
Less than $10,000	302	239	541
$10,000 or more	194	240	434
Total	496	479	975

Source: ICPSR (1977).

Table 4.1 is a cross-tabulation of type of occupation (manual or non-manual) and income (for the sake of simplicity, divided into just two categories) from a random sample of US citizens. Is there a relationship between the two variables? To answer this question, we need first to be clear about what is meant by "relationship". There are several equivalent ways of thinking about the relationship between two variables. One is to see whether respondents are distributed across the categories of one variable in the same proportions, regardless of which category of the other variable they are in. Table 4.1 shows that the distribution of respondents in the two income categories differs between the two occupation categories. There are proportionately more manual than non-manual workers in the lower-income category. We can be more precise about this by drawing up a percentaged table like

Table 4.2. This shows that 61 per cent of manual workers earned less than $10,000, as compared with only 50 per cent of non-manual workers. Because the distribution of respondents between income categories differs between the occupation categories, we say that there is a relationship, specifically an association, between the variables occupation and income in the sample.

Table 4.2 Percentage distribution of income, by occupation.

Income	Occupation (%)		
	Manual	Non-manual	Total
Less than $10,000	61	50	56
$10,000 or more	39	50	54
Total	100	100	100
	(496)	(479)	(975)

Another, equivalent, way of thinking about association is in terms of *probabilities*. There is a 61 per cent probability of a manual worker from the sample having an income below $10,000, but only a 50 per cent probability for a non-manual worker. Thus, amongst the sample, a respondent's occupation does make a difference to the probability that he or she has an income below $10,000, and this shows that there is association. These two ways of viewing association are statistically equivalent. Both show the existence of association between income and occupation in these data.

There is another, quicker, way of telling whether there is association in a table of two variables when each variable has only two categories. (Such tables are called "2 by 2" tables.) Multiply the top left-hand frequency and the bottom right-hand frequency, and divide the result by the product of the top right-hand and bottom left-hand frequencies. This gives a number known as the *odds ratio*, which is equal to 1 if there is no association. The odds ratio for Table 4.1 is:

$$\frac{302 \times 240}{239 \times 194} = 1.56$$

so the variables are associated. Sometimes the odds ratio is called the *cross-product ratio*. The name "odds ratio" is taken from gambling jargon: the odds of being a manual worker earning less than $10,000 is 302 to 194 and the odds of a non-manual worker earning less than $10,000 is 239 to 240, so the odds ratio is

$$\frac{302/194}{239/240} = \frac{302 \times 240}{239 \times 194} = 1.56$$

Although there is association in this table, its strength (the amount of association) is quite modest. The probability of earning less than $10,000 changes by only 11 per cent from the manual to non-manual category. A reason for the low degree of association can be seen in Tables 4.3 and 4.4, in which the data have been divided into two separate tables, one for men and the other for women. Tables 4.3 and 4.4 are two partial tables derived from Table 4.1, controlling for the gender of the respondent. (Gender is the *control variable*). Since all the respondents have been placed in one or other of the two partial tables, the cell-by-cell sums of the entries in these tables equal the entries in the corresponding cells of the full table, Table 4.1.

Table 4.3 Income, by occupation, for male respondents only.

	Occupation		
Income	Manual	Non-manual	Total
Less than $10,000	148	51	199
	(46%)	(25%)	
$10,000 or more	175	157	332
	(54%)	(75%)	
Total	323	208	531
	(100%)	(100%)	

Table 4.4 Income, by occupation, for female respondents only.

	Occupation		
Income	Manual	Non-manual	Total
Less than $10,000	154	188	342
	(89%)	(69%)	
$10,000 or more	19	83	102
	(11%)	(31%)	
Total	173	271	444
	(100%)	(100%)	

The first thing to notice about these partial tables is that each shows a much greater degree of association between income and occupation than the full table. In other words, the differences between the distributions of income for manual and non-manual workers, keeping the men and women separate, is greater than the difference for all workers combined. Again, this is easiest to see by looking at the percentage figures. Secondly, while non-manual males tend to be paid over $10,000, non-manual females are relatively much worse paid. Conversely, a higher proportion of males than females in manual jobs earns $10,000 or more. This means that the associations in the two partial tables are quite different and are almost mirror-images of each other. Summing the two partial tables to form the full table tends to "wash out" or mask the opposing strong associations visible in the single-sex partial tables, and this is why quite a low level of association was found in the full table (Table 4.1).

To summarize, there is some association in all three tables, but the association between occupation and income is affected by the gender of the respondent. We conclude, therefore, that gender is an important factor in determining the association between occupation and income in the sample. Women tend to be paid less than men for doing jobs of the same kind. To distinguish the associations in these tables, those in the partial tables are called the *partial associations*, and the association in the full table is the *marginal association* ("marginal" because this table is a marginal table of the three-dimensional income by occupation by gender table).

We noted above that, when two variables such as occupation and income are related, so that the level of one variable makes a difference to the distribution of respondents on the other, the relationship is termed "association". When three variables are related, in a way such that the association between two of them changes according to the level of the third, the relationship between the three is called *interaction*. (This statistical meaning of "interaction" has nothing to do with the other common use of the word, in sociology and social psychology, to refer to exchanges between two or more people.) Since the association between income and occupation is quite different for men and for women, there must be some interaction between income, occupation and gender. Sometimes interaction is called "specification" (Rosenberg 1968) because, for example, gender helps to specify the degree of association between income and occupation.

The interaction we have found can be interpreted as follows. Women tend to obtain low-paid non-manual jobs, becoming clerks, typists, nurses and shop assistants, rather than taking up the managerial and professional posts filled by men. It is to be expected within a capitalist system that income and occupation would be associated, but the difference in the partial associations according to gender is a clear indication of women's lack of equality of income. If in future the amount of interaction were to get smaller, this would be a sign of increasing sexual equality.

The procedure used in this example, first examining a cross-tabulation of two variables, then looking for changes in the degree of association between the two variables after controlling for a third, is called *elaboration*. Elaboration is a simple technique for analyzing the mutual influence of three variables when one (for instance, gender) is clearly causally prior to (that is, has a causal effect on) the other two, and it has been much used by social researchers (Rosenberg 1968). However, in practice it is limited to the analysis of three cross-classified variables. Since a partial table is obtained for each category of the control variable, if this has more than two or three levels the comparison of partial tables quickly becomes unmanageable. Moreover, in many cases it is not as clear as in the example which of the three variables is best chosen as the control variable. It is only because we know that gender may affect income and occupation, but income and occupation cannot affect gender, that it was easy to see that gender must be the variable to control.

Cross-tabulations of more than three variables, such as the table of experience of pre- and extramarital sex in Chapter 3, cannot conveniently be analyzed using elaboration, for there are just too many partial tables needing to be compared. But the most significant failing of the elaboration technique is that the analysis can proceed without developing a clear model of the interrelationships between the variables. To put it bluntly, elaboration allows you to muddle through, picking on one or other relationship to examine, and does not force you to think clearly about what you expect to find in your data. In contrast, the loglinear techniques to be introduced in the next chapter require the formulation of models from prior theoretical expectations. These models can be developed in a systematic way to explore all the relationships in the data.

Analyzing a two-dimensional table

To show how a model might be used to analyze data, we shall return to the simplest relationship, an association between two variables, and employ some of the data on sexual experiences introduced in Chapter 3 for a further example. This will concentrate on the principles of the analysis in order to lay a foundation for the more complex examples to be discussed later. Table 4.5 is one marginal from the full table, Table 3.2, cross-tabulating reports of experiences of premarital sex with current marital status. A model will be used to answer two questions: are the two variables related, and, if so, how strong is the relationship?

Table 4.5 Frequency of reporting experience of premarital sex (PMS), by marital status.

	Marital status		
Reporting PMS	Married	Divorced	Total
No	460	335	795
Yes	82	159	241
Total	542	494	1036

The first step of the analysis is to choose a model that expresses the presumed relationship (or lack of it) between the two variables. We wish to find the simplest model that will satisfactorily represent the relationship, so we begin by seeing whether a model that involves no association between marital status and reports of premarital sex is an adequate one. This means that we should try to fit a "model of no association". If the model generates data similar to the data that have been collected, then we may conclude that there is indeed no relationship between the two variables. On the other hand, if the modelled data do not resemble the real data, we must suppose that the likelihood of reporting experiences of premarital sex depends on marital status.

In terms of the ideas introduced in Chapter 1, we compare the "real world" represented by the data of Table 4.5 with an "imaginary world" represented by a table obtained using a model of no association. If the tables are identical, there is no association between the variables; if they differ, the model does not fit and the variables are associated.

Thus, the second step of the analysis is to construct a table, the model table, showing the distribution of respondents if there were no association between marital status and reporting premarital sex. The table must be identical to the data in all possible respects, other than in having no association. This is necessary in order to ensure that the "real" and "imaginary" worlds are the same excepting only those relationships specified in the model. In particular, the model table must include the same total number of respondents, and the same proportions of respondents must be married and must report premarital sex. In short, the marginals of the model table must equal the marginals of the data table. The body of the model table must be filled with those frequencies that would be obtained if there were no association, since this is how we wish the "imaginary world" to be.

The required frequencies can be calculated very simply by multiplying together the marginal entries to the right and below each cell, and dividing by the table total. For instance, the frequency for the top left cell of the model table (married; no premarital sex) must be 795 multiplied by 542 divided by 1036. Table 4.6(a) shows the results of these calculations. This table includes the frequencies that would be expected in an "imaginary world" where there was no association between marital status and reporting experience of premarital sex. As a check that it does indeed show no association between the variables, the percentages across the table can be computed to demonstrate that the distribution of cases is the same for the two levels of premarital sex. These are shown in part (b) of Table 4.6.

We now have the frequencies to be expected in an "imaginary world", to compare with the frequencies which were actually collected in the "real world". If the data table (Table 4.5) and the model table were identical, the conclusion to be drawn would be that the two variables are not related. However, the two tables are clearly different, and this means that there is some association. The initial model of no association is revealed to be incorrect. We can infer that being divorced and reporting one's premarital adventures tend, in the "real world", to go together.

Table 4.6 Table for model of no association.

(a) Frequencies

	Marital status		
Reporting PMS	Married	Divorced	Total
No	416	379	795
Yes	126	115	241
Total	542	494	1,036

(b) Percentaged across marital status, to show no association

	Marital status (%)		
Reporting PMS	Married	Divorced	Total
No	52.3	47.7	100
Yes	52.3	47.7	100

The method used to calculate the cell entries in the model table (also called the *expected* or *fitted* table) can be expressed as

model cell entry = (product of corresponding marginals)/table total

which in the generalized notation introduced in Chapter 3 becomes

$$m_{ij} = x_{i+}x_{+j} / x_{++}$$

where m_{ij} stands for the frequency in cell i,j of the model table and x_{ij} the frequency in the corresponding data table cell. Recall that the plus sign means sum over cells, so that, for instance x_{i+} is the marginal obtained by summing over all values of j.

So far we have concluded only that, since the model and data tables differ, there is some association in the data. We may also be interested in the strength of this association. We have already met one measure of association, the odds ratio, but this has the disadvantage that its value can range from zero to infinity according to the amount of association in the table. A number of other measures have been designed to have the more convenient range of 0–1, 0 meaning no association and 1, perfect association. Several are in common use, each with its own particular merits (see Goodman & Kruskal 1954), but the most often used for 2×2 tables is named after the Greek letter *phi*, (ϕ). This is computed from the formula:

$$\phi^2 = \frac{\sum \left(x_{ij} - m_{ij}\right)^2 / m_{ij}}{\sum m_{ij}}$$

where Σ means "form the sum of the expression following it for all combinations of values of i and j", or, more concisely, sum the expression calculated for each cell in the table. The formula above gives the value of ϕ^2. ϕ itself is, of course, obtained by taking the square root.

ϕ can be seen from the formula to be based on the magnitude of the cell-by-cell differences between the data and the model table (that is, $x_{ij} - m_{ij}$). It can be used only on two-dimensional tables in which both variables have only two levels. It is thus restricted to measuring association in simple tables of four cells. A related measure, Cramer's V, is used for two-dimensional tables in which one or both of the variables have more than two levels.

Let us work through an example to illustrate the meaning of the formula for ϕ. First, taking each cell in turn, we subtract the model table cell entry (m_{ij}) from the data table cell entry (x_{ij}), square the result, and divide by the model table cell entry. For the data shown in Table 4.5, the top left-hand cell entry (x_{ij}) is 460. The corresponding model table cell entry (m_{ij}), from Table 4.6, is 416. Subtracting 416 from 460 gives 44; squaring and dividing by 416 gives 4.65. Similar calculations carried out for the other three cells yield, for the top right cell, 5.11; for the bottom left cell, 15.4; and for the bottom right cell, 16.8. Next, the summation sign in the numerator of the formula indicates that these four numbers need to be added, giving 42.0. Finally, we divide by the total obtained by summing all four model table cell entries (1036) to give ϕ^2, 0.0404, and a value for ϕ itself of 0.201, representing a modest level of association.

Although we have found some association between reporting premarital sex experiences and marital status in this data table, we cannot jump immediately to the conclusion that there is such an association in the population at large. The calculation of ϕ has only used data from a sample of 1,036 people. It is possible that, for instance, the randomly chosen individuals who were asked about their experiences happened to include an unrepresentatively large number of promiscuous divorcees. The probability of the data being unrepresentative by chance in such a way that we could falsely find an association when none really existed can be assessed using a test of significance. A measure of significance is also a valuable guide to the fit of models when there are a number of alternatives to select from. In the simple case of a two-dimensional table, however, the only interesting models to examine are those of "no association" and its converse, the model of association. If we can reject the no-association model, it follows that the alternative association model is the one to choose.

Chi square (χ^2) is the most common measure of significance (Caulcott 1973). In the generalized notation, its formula is:

$$\chi^2 = \sum (x_{ij} - m_{ij})^2 \; / \; m_{ij}$$

This is just the numerator of the formula for ϕ^2. The interpretation to be placed on the χ^2 value depends on a characteristic of the model we are testing, called its *degrees of freedom*. The degrees of freedom of a model is the inverse of the number of restrictions or constraints that have been imposed on the model table to make it conform to the data. The degrees of freedom measures the number of ways in which the "imaginary world" is permitted to vary from the "real world".

To begin with, there are as many degrees of freedom as there are cells in the table. But in defining the model table, we imposed three constraints: first, that the overall total number of respondents is fixed to be equal to the number in the data (1036); secondly, that the marginal of the premarital sex variable is distributed identically in both model and data tables; and thirdly, that the marital status marginal is distributed identically in both tables. The imposition of each of these constraints reduces the model table's degrees of freedom by one. Since Table 4.6 contains four cells and we have imposed three constraints, it follows that the model has one degree of freedom.

Returning now to the comparison of the data with the "no-association" model, application of the formula above for the value of χ^2 gives:

$$\chi^2 = 42.1$$

To see what this means, it must be assessed against the theoretical distribution of χ^2, which can be found in standard reference books and in the appendices of many statistics textbooks, or can be computed by a data analysis program. The table of the χ^2 distribution for one degree of freedom indicates that the probability of obtaining a value as high as 42.1 if the variables are not associated is exceedingly small. Putting it another way, if premarital sex and marital status were independent, the chances of selecting a random sample that would result in a χ^2 value of 41.1 is less than 1 in 10,000. Because it is so improbable that we would get such a χ^2 value if there is no association in the population, we can safely conclude that pre-marital sex and marital status are associated. We can be confident that we are right in rejecting the no-association model in favour of one that includes association.

Summary

Two variables are associated if the proportion of respondents in the categories of one variable differs between the categories of the other. Three variables interact if the degree of association between two of them differs between the categories of the third. Three-variable (three-dimensional) tables may be analyzed using *elaboration*, which involves comparing the partial tables relating two variables while controlling for the third, but the technique becomes clumsy if the variables have more than two levels. A better method of analyzing three- and higher-dimensional tables is loglinear analysis, which requires the specification of explicit models.

The analysis of a table using a model was illustrated with an example of a simple two-dimensional table. The logic lying behind this example is important, because the same logic is used in the much more complicated case of tables with three or more variables. An observed data table was compared with another table constructed on the assumption of a specified simple model. The model was arranged to mirror the data in all respects, except that it did not include any association between the variables. This was achieved by ensuring that the model table's marginals were kept the same as those of the data table. An inferential test, the χ^2 test, was used to decide whether the model fitted the data. The magnitude of the association was measured with ϕ, which expressed the extent of the deviation of the observed table from the model table. We shall follow essentially the same course of argument in Chapter 5 when analyzing three-dimensional tables using loglinear analysis.

Computer analysis

SPSS

SPSS will calculate many different measures of association on request, some of them rather obscure, others used only in special circumstances. The STATISTICS subcommand of the CROSSTABS procedure is used to specify which measures are to be calculated. Program 4.1 will read in the income by occupation by gender data of Table 4.3, display the marginal tables of income by occupation and the two partial tables for each gender separately, and calculate the χ^2 and ϕ statistics for all three tables.

Program 4.1

```
DATA LIST FREE / Income Occup Gender Count.
BEGIN DATA.
1 1 1   148
1 1 2   154
1 2 1    51
1 2 2   188
2 1 1   175
2 1 2    19
2 2 1   157
2 2 2    83
END DATA.
WEIGHT BY Count.
CROSSTABS Income BY Occup
   / Income BY Occup BY Gender
   /CELLS = COUNT, COLUMN
   /STATISTICS = CHISQ, PHI.
```

GLIM

GLIM does not include built-in facilities for calculating measures of association, although it is possible to write macros to calculate them. For example, a GLIM program can be written to use the formula above to compute the value of ϕ for a table.

That is what Program 4.2 does: it reads the data of Table 4.1, computes ϕ and χ^2 and prints them.

Program 4.2

```
$units 4
$data Income Occup Count
$read
1 1   302
1 1   239
1 2   194
1 2   240
$yvar Count
$error Poisson
$fit Income + Occup
$calculate %a = %sqr(%cu((Count - %fv)**2/%fv)/%cu(%fv))
$print 'Phi coefficient = ' %a
$print 'Chi square = ' %x2
```

This program reads the data, fits a model of no association, calculates ϕ using the system scalar %fv, which contains the *fitted values*, and prints ϕ and χ^2 (which is available after a model has been fitted in the system scalar, %x2).

The $yvar directive tells GLIM that it is the cell counts in the variable Count that are to be modelled. ("yvar" stands for "*y* variate", following the convention that the letter *y* is used for the variable to be explained or modelled.) Because the cell counts come from a random sample, the precise distribution of counts into cells will differ randomly between samples, following a "Poisson" distribution. This is specified to GLIM in the $error directive (see Chapter 10).

If you want to calculate ϕ frequently, typing the long $calculate directive again and again is tedious and prone to error. It is better to put the calculation into a *macro*, storing it separately in a file so that it can be used repeatedly by specifying its name. Program 4.3 is a macro called PHISTATS, which prints ϕ and χ^2 and also computes the χ^2 probability to save looking it up in tables. The calculation of the probability level uses a complex formula which yields a good approximation to the correct values.

Program 4.3

```
$MACRO PHISTATS
$c Calculates and prints the phi statistic and chi square
$calculate %a = %sqr(%cu((%1- %fv)**2/%fv)/%cu(%fv))
$calculate %p = %eq(%df,1)*(2-2*%np(%sqrt(%dv)))+
   %eq(%df,2)*(%exp(-%dv/2)) +
   %gt(%df,2)*(1-%np(((%dv/%df)**(1/3)-
   1+2/(9*%df))/%sqrt(2/(9*%df))))
$print 'Phi coefficient = ' %a
$print 'Chi square = ' %x2
$print 'Probability level =' %p $
$endmac
```

Once you have typed this macro, it can be executed with the directive $use PHISTATS Count. Program 4.4 provides an example which carries out the same analysis as above. Typing has been reduced still further by abbreviating "Poisson" in the $error directive to just "p".

Program 4.4

```
$units 4
$data Income Occup Count
$read
1 1   302
1 1   239
1 2   194
1 2   240
$yvar Count
$error p
$fit Income + Occup
$use PHISTATS Count
```

Further reading

The concept of association between variables is treated in most texts on social statistics. Loether & McTavish (1992, P. III) is especially recommended; Part IV also explains elaboration and interaction clearly. Longer presentations of the method of elaboration are to be found in Rosenberg (1968) and Babbie (1973). Caulcott (1973) provides a very thorough explanation of χ^2 values. Blalock (1979), in Appendix 1, includes a short "refresher" on the algebra of summations and logarithms for those whose mathematics is a little rusty.

CHAPTER FIVE
Loglinear analysis

As we saw in Chapter 4, modelling a two-variable table is fairly easy. As soon as more variables are introduced, however, there are many more relationships to be considered. In a three-dimensional table there may be associations among each of the pairs of variables as well as interaction between all of them. In four- and five-dimensional tables the number of possible relationships multiples alarmingly. Fortunately, once you understand how to use loglinear models with three-dimensional tables, extension of the technique to more complex ones is straightforward.

Table 5.1 is a three-dimensional table relating the variables occupational class, tenure and voting among a sample of men in Banbury in

Table 5.1 Occupational class, tenure and voting among economically active men in Banbury and district, 6 per cent sample, 1967.

Occupational class	Own outright	Own on mortgage	Rented privately	Rented from council
(a) Voted Conservative				
I and II	35	63	22	5
IIIa, IVa	18	29	18	16
IIIb	12	30	16	24
IVb	9	12	8	17
V	5	1	5	3
(b) Voted Labour				
I and II	5	14	3	10
IIIa, IVa	4	9	4	14
IIIb	3	44	83	42
IVb	7	18	14	45
V	0	2	5	21

Source: derived from Stacey et al. (1975: Table 4.1)

1967. The respondents classified in this table were asked to state their occupation (measured on a version of the Registrar-General's scale and used as an indicator of class), whether and how they owned or rented their home (tenure), and how they voted in the last election.

As Chapter 4 mentioned, interaction is said to occur when the degree of association between two variables differs between the categories of a third. For instance, there is interaction in Table 5.1 if the magnitude of the association between class and tenure among those voting Conservative (the top half of the table) is not the same as the magnitude of the association between these two variables among those voting Labour (the bottom half). This is the meaning of interaction in statistical terms. But how can it be interpreted sociologically? The authors of the report in which the table was published write that:

> Property ownership, including domestic property, and the work situation have, since Marx and Weber, been thought of as bases for class formation and for political action. Voting at the last election may be taken as one indicator of political belief and action. [Table 5.1] suggests that there does appear to be a relationship between property ownership, higher occupational status and Conservative voting . . . Those who own outright are the most Conservative category and include over 80 per cent in all classes from skilled manual and above. Among those who rent privately or who own subject to mortgage it is only in the non-manual categories that there is a large Conservative majority. Those who rent from the Council are most inclined to vote Labour, and this is true even if they are in non-manual categories. It seems, therefore, that house tenure and occupational status combined account for more of the variation in voting than either taken separately. (Stacey et al. 1975: 41)

The suggestion in the last sentence is that Table 5.1 includes statistical interaction. Voting behaviour depends on type of tenure and class. But, in addition, the magnitude of the relationship between voting and tenure is affected by class. It is the variation in association between tenure and vote according to class that is meant when it is said that the table displays interaction. The consequence of this interaction is that the probability of voting Conservative depends not only on the independent effects of type of home tenure and class position but also on the effect of these two influences in combination.

In fact, if there is interaction, it will be found that, not only does the association between tenure and vote depend on class, but also the association between class and vote depends on tenure, and the association between class and tenure depends on vote. This is because interaction is a symmetrical property: if the association between two variables depends on a third, the association between each of the other pairs of variables will also depend on the remaining variable.

Loglinear analysis will be introduced by discovering whether there is interaction in Table 5.1. We shall construct a model table which shows no interaction and then compare this table with the Banbury data. If the model and data tables are similar, we shall have developed a simpler model than the original study assumed was necessary. Moreover, we shall have shown that there is in fact no interaction in the data and that the study was wrong in arguing that occupational class and tenure have a joint effect on voting behaviour over and above their separate influences.

In Chapter 4, a model of "no-association" was constructed by fixing the marginals to be the same in both data and model tables. The same can be done to calculate a model to compare with Table 5.1. However, with three-dimensional tables, there are not only the marginals for each variable on its own, but also marginal tables showing the relationships between pairs of variables. The three marginal tables that can be derived from Table 5.1, for the pairs of variables [class, vote], [tenure, vote] and [class, tenure], are shown in Table 5.2.

Table 5.2 Three marginal tables from the Banbury data of Table 5.1.

	Vote	
Occupational class	Conservative	Labour
(a) Marginal table of occupational class, by vote		
I and II	125	32
IIIa, IVa	81	31
IIIb	82	172
IVb	46	84
V	14	28
(b) Marginal table of tenure, by vote		
Own outright	79	19
Own on a mortgage	135	87
Rented privately	69	109
Rented from council	65	132

(c) Marginal table of occupational class by tenure

	Tenure			
Occupational class	Own outright	Own on mortgage	Rented privately	Rented from council
I and II	40	77	25	15
IIIa, IVa	22	38	22	30
IIIb	15	74	99	66
IVb	16	30	22	62
V	5	3	10	24

We now need to compute model table frequencies such that the model table has the same marginal tables as the data marginal tables of Table 5.2. This was an easy step for the model of "no association" calculated in Chapter 4, since it merely involved applying the formula

$$m_{ij} = x_{i+}x_{+j} / x_{++}$$

for each model table cell in turn. Unfortunately, calculation of model table frequencies is not so straightforward when dealing with a model of "no interaction", since a standard formula does not exist in this case. However, although there is no formula, there are several methods or algorithms which can be used. The easiest of these algorithms to understand is known as *iterative proportional scaling*. This procedure can best be explained by an analogy that will be familiar if you do household repairs.

One way to mix the kind of plaster used for filling cracks in walls is to obey the instructions on the packet which specify how much water is to be added to how much powder. But the way most people do it is to add a little water to some powder, stir the mixture and test to see whether the result has the right consistency. If it is too dry, some more water is added, or if it is too wet, some more powder. Then the mixture is stirred and tested again. This goes on until the balance of powder to water seems right. The advantage of this method is that you are more or less bound to get to the correct mixture eventually, provided only that you add a moderate and decreasing amount of water or powder at each stage.

Making plaster like this is an example of an *iterative* procedure. Iterative procedures are also useful in data analysis, both for finding model table frequencies and for fitting other kinds of complex models. The essential elements of an iterative procedure are as follows:

(1) A guess at a solution is made (for example, some powder and a little water).

(2) The guess is tested to see whether it is good enough (the mixture's consistency is checked), and if it is, this guess is the solution.
(3) The guess is improved (a little more water or powder is added).
(4) Steps 2–4 are repeated until a satisfactory solution is discovered in step 2.

The great advantage of iterative methods is that they can be used even when there is no exact formula known for working out the result.

In the case of fitting models, a series of guesses of the model table counts are made, matching in turn each marginal, which should be the same in the data and the model table. Each adjustment to take account of the constraint of one marginal will disturb the match of the others, but by decreasing amounts each time until eventually all the marginals to be fitted match the data.

For example, to calculate the frequencies for a model of "no interaction", a guess is first made at a solution. For simplicity of calculation, this first guess is usually a table in which every cell frequency is equal to one. It does not matter that this first guess is almost certainly a very poor one; the iterative method will soon improve it. That was step 1: make a guess at a solution. Next, we must test the solution to see whether it is adequate (step 2). The test is whether the marginals of the guessed table are the same as those in the data table. If they are, the desired model table has been found; if not, the guess must be improved (step 3).

Progress so far is shown in part (a) of Table 5.3. The right-hand side of the table is the class-by-tenure marginal from the first guess. It is to be compared with the corresponding data marginal shown below it. Since the two marginals are clearly quite different, solution (a) is not satisfactory and must be improved. The improvement is performed by proportionately scaling the frequencies, that is, multiplying each of them by the ratio of the data to solution marginal table entries. For instance, the second-guess table's top right frequency is

$$\frac{[\text{first-guess table's frequency}] \times [\text{data marginal entry}]}{[\text{first-guess marginal entry}]} = \frac{1 \times 40}{2}$$
$$= 20.0$$

Carrying out the scaling for all the table frequencies yields the solution in (b), the second guess. The scaling formula ensures that the class-by-tenure marginal table from the second guess is exactly equal to the data marginal, as desired. However, the other two marginals may not and

Table 5.3 Iterative proportional scaling on the data in Table 5.1.
The data table (Table 5.1 rearranged, with the labels omitted to save space)

35.0	63.0	22.0	5.0	5.00	14.0	3.0	10.0
18.0	29.0	18.0	16.0	4.00	9.0	4.0	14.0
12.0	30.0	16.0	24.0	3.00	44.0	83.0	42.0
9.0	12.0	8.0	17.0	7.00	18.0	14.0	45.0
5.0	1.0	5.0	3.0	0.00	2.0	5.0	21.0

(a) First guess at a solution

								Class by tenure marginal							
								From first guess				From data			
1	1	1	1	1	1	1	1	2	2	2	2	40.0	77.0	25.0	15.0
1	1	1	1	1	1	1	1	2	2	2	2	22.0	38.0	22.0	30.0
1	1	1	1	1	1	1	1	2	2	2	2	15.0	74.0	99.0	66.0
1	1	1	1	1	1	1	1	2	2	2	2	16.0	30.0	22.0	62.0
1	1	1	1	1	1	1	1	2	2	2	2	5.0	3.0	10.0	24.0

(b) Second guess at a solution

								Class by vote marginal			
								From second guess		From data	
20.0	38.5	12.5	7.5	20.0	38.5	12.5	7.5	78.5	78.5	125	32.0
11.0	19.0	11.0	15.0	11.0	17.0	11.0	15.0	56.0	56.0	81.0	31.0
7.5	37.0	49.5	33.0	7.5	37.0	49.5	33.0	127	127	82.0	172
8.0	15.0	11.0	31.0	8.0	15.0	11.0	31.0	65.0	65.0	46.0	84.0
2.5	1.5	5.0	12.0	2.5	1.5	5.0	12.0	21.0	21.0	14.0	28.0

(c) Third guess at a solution

								Tenure by vote marginal			
								From third guess		From data	
31.8	61.3	19.9	11.9	8.15	15.7	5.10	3.06	59.9	38.1	79.0	19.0
15.9	27.5	15.9	21.7	6.09	10.5	6.09	8.30	24	97.7	135	87.0
4.84	23.9	32.0	21.3	10.2	50.1	67.0	44.7	78.9	99.1	69.0	109
5.66	10.6	7.78	21.9	10.3	19.4	14.2	40.1	84.8	112	65.0	132
1.67	1.0	3.33	8.0	3.33	2.0	6.67	16.0				

(d) Fourth guess at a solution

42.0	66.6	17.4	9.15	4.07	14.0	5.60	3.60
21.0	29.8	13.9	16.6	3.04	9.37	6.70	9.78
6.38	25.9	28.0	16.3	5.07	44.6	73.7	52.6
7.46	11.5	6.81	16.8	5.16	17.3	15.6	47.2
2.20	1.09	2.92	6.13	1.66	1.78	7.33	18.8

probably will not be correct. The first guess was checked and improved with the class-by-tenure marginal; the second guess is improved in the same way, but using the class-by-vote marginal to yield the third guess (part (c) in the Table). The third guess is compared with the remaining marginal (tenure-by-vote); the fourth (part (d)) with the class-by-tenure marginal like the first; and so on. The iterations continue until all three marginals from the solution match the marginals from the data, either precisely or sufficiently closely for any differences to be unimportant. The order in which the marginal tables are used in the calculation makes no difference to the result. The iterative scaling procedure almost always converges to a solution, usually within three or four cycles. It may seem laborious, but in practice it is always carried out by computer, so the amount of calculation does not matter much.

The model table obtained by continuing the process begun in Table 5.3 is shown in Table 5.4. This table has the same marginals as the data table (Table 5.1) and is the table of frequencies that would be obtained if there were no interaction between class, tenure and vote. Tables 5.1 (the data) and 5.4 (from the model) are different, although there is a certain similarity in the cell frequencies. This shows that the original conclusion, that there *is* interaction, was correct (although the interaction is not very strong). Later, we shall be able to quantify the amount of interaction, but for the moment it can be said that class and

Table 5.4 Model table for model of "no-interaction" fitted to Banbury data.

Occupational class	Own, outright	Own, on mortgage	Rented, privately	Rented, from council
Voted Conservative				
I and II	36.3	61.7	17.4	9.64
IIIa, IVa	19.6	29.2	14.4	17.9
IIIb	9.36	29.8	27.5	15.3
IVb	10.4	13.0	6.68	15.9
V	3.30	1.32	3.10	6.28
Voted Labour				
I and II	3.68	15.3	7.61	5.36
IIIa, IVa	2.42	8.83	7.65	12.1
IIIb	5.64	44.2	71.5	50.7
IVb	5.55	17.0	15.3	46.1
V	1.70	1.68	6.90	17.7

type of tenure together have some, but not much, more effect on voting behaviour than either considered separately.

Now that we have a model table, it can be used to provide a numerical illustration of the effect of interaction. For clarity, let us look at only one small portion of the table, i.e. those cells concerned with people who rent either privately or from the council. For the "top" classes, I and II, the association between the kind of landlord (private or council) and vote can be measured by the odds ratio for the "2 × 2" subtable with cell entries: 17.4 and 9.64 (the Conservatives), and 7.61 and 5.36 (the Labour voters):

$$\frac{17.4 \times 5.36}{9.64 \times 7.61} = 1.27$$

The cell frequencies for the four corresponding cells for class V (3.10, 6.28, 6.90 and 17.7) can be slotted into another "2 × 2" table, showing the association between kind of landlord and vote amongst class V respondents. The odds ratio for this table is:

$$\frac{3.10 \times 17.7}{6.28 \times 6.90} = 1.27$$

The two odds ratios are exactly the same because there is no interaction in the model table. Repeating these calculations for the frequencies in the data table gives an odds ratio of 14.7 for classes I and II, and 7 for class V, the two values differing because the data table includes some interaction.

In this example, we developed a model table starting from the assumption that class, tenure and vote were associated, but did not interact. In the next illustration, the most obvious model to examine is one that involves not three but only two associations, again with no interaction. Table 5.5, a cross-tabulation of occupational class, gender and voting behaviour, is also taken from the Banbury study. Occupational class has been divided in this table into only two categories, manual and non-manual, since this is how the data were presented in Stacey (1975).

The previous example showed, if indeed further evidence is needed, that one's class influences the way one votes. Gender and political persuasion are also related, women having traditionally been more likely to vote Conservative, regardless of their class position. A model to fit the data in Table 5.5 would therefore need to allow for associations between class and vote, and gender and vote. Note that each of these associations is presumed to exist independently of the influence of the third variable. For instance, to say that class and vote are associated means that men's voting behaviour depends on their class position, and likewise that women's voting depends on their class position.

Table 5.5 Voting Labour or Conservative, by gender and occupational class, for Banbury and district, 6 per cent sample, 1967.

Occupational class	Vote*			
	Conservative		Labour	
	Male	Female†	Male	Female†
Non-manual	140	152	50	50
Manual	109	136	215	159

* Respondents voting Liberal (52, 6% of total sample of 891) or not voting (220, 25% of sample) excluded from table.
† Single women classified by their own occupation, those ever married by their husbands' occupation.
Source: derived from Stacey et al., (1975, Table XV).

It is more difficult to decide whether to include a relationship between class and gender in the model. The issue is complicated by Stacey's decision to categorize all married women, employed or not, by the occupation of their husbands. Since the majority of the female respondents (about 77 per cent) were married, most women were actually classified by their husbands' occupations. The effect of this is to reduce considerably any possible relationship between class and gender. It is, therefore, quite likely that there is no association between these variables in Table 5.5.

These considerations, together with the aim of using the simplest possible model that fits the data acceptably well, suggest that we should first try a model in which only class and vote, and gender and vote, are associated. In this model, class is independent of gender among the Conservatives, and the same is true among the Labour voters. The necessary model table frequencies can be computed using iterative proportional scaling.

The model table has to be designed so that the two associations to be included are fixed to be the same as those in the data. This is achieved by ensuring that the marginal tables for class by vote and for gender by vote are the same in both data and model tables. Likewise, the total number of respondents and the marginals for each of the three variables must be the same in both data and model tables. However, the association between class and gender in the model table must not be the same as in the data; on the contrary, the frequencies in the model table must be arranged to give zero association between these variables in each (Conservative and Labour) partial table.

Because only two of the marginal tables are required to be the same for the data and the model tables, the scaling is done with respect to these two only. First, the guessed solution is scaled using the class-by-vote marginal, then the resulting second guess is scaled using the gender-by-vote marginal, the third guess is scaled using the class-by-vote table again, and so on, until the model table's class-by-vote and gender-by-vote marginals equal those of the data table. Because the class-by-gender marginal was not used in the scaling, the iterative procedure automatically yields frequencies that show no partial associations between this pair of variables.

The result of iterative scaling in this way is shown in Table 5.6. Neither of the partial tables of class-by-gender controlling for vote display any association, as required by the model specification.

Table 5.6 Model table for vote, by class and gender, for a model of no interaction and no association between class and gender.

Occupational class	Vote			
	Conservative		Labour	
	Male	Female	Male	Female
Non-manual	135	157	56	44
Manual	114	131	209	165

It is important to note that the fact that the partial tables for class-by-gender, controlling for vote, show no association does not mean that the class-by-gender marginal table will also show no association. The amount of marginal association depends not only on the partial associations but also on the other associations in the table.

Table 5.6, based on a model of no-interaction and no-association between class and gender, can be compared with the data of Table 5.5. The two tables do differ, but not by a great deal. It is not immediately clear whether the model table is sufficiently similar to the data table to allow the conclusion that the model is adequate. This judgement must be made by calculating a measure of the difference between the tables like the χ^2 statistic used in Chapter 4. Such a statistic will be introduced in Chapter 6 after a few more features of loglinear analysis have been discussed. If you were to use it now you would find that, according to the conventional criterion, the model does fit the data adequately. It can be inferred that there is no significant association between class and gender in the data, just as we had initially supposed.

One further point needs to be made before leaving this example. It was assumed when we set up the model that there is an association between gender and vote. This might have been incorrect, but, even so, the model would still have fitted the data. The model was constructed to mirror precisely the association between these two variables, no matter how small or large that association might be. If in fact the association in the data had been zero, then that too would have been reproduced in the model. Consequently, the fact that an association between gender and vote was included in the model, and that the model fits, does not prove that this association is present.

We could next try fitting another model, one in which the associations between gender and vote, as well as between class and gender, were explicitly omitted. Indeed, this would be a useful next step. If the reduced model seemed to fit, we could try removing the class-by-vote relationship. In this way, we can work towards the simplest model that fits the data, omitting unnecessary relationships. Later it will be seen how the process of model selection can be systematized, but first there is a more thorough discussion of how to judge whether one model fits better than another.

Summary

Loglinear analysis is used to construct a table that includes only those relationships specified in a model. The resulting table can then be compared with the data table to see whether the model is a good one. One way of calculating model table frequencies is to use a method known as iterative proportional scaling. This method generates a table in which some specified marginals are fixed to be identical to those in the data table, while others are left unconstrained. The fixed marginals are those that correspond to the relationships included in the model to be examined.

Computer analysis

The following programs show how to carry out a basic loglinear analysis using SPSS and GLIM.

SPSS

SPSS provides two entirely separate commands in its Advanced Statistics Module for carrying out loglinear analyses, HILOGLINEAR and LOGLINEAR. The former uses the iterative proportional scaling algorithm introduced in this chapter, while the latter uses a more flexible but more complex algorithm. It is usually best to use the HILOGLINEAR command unless you need features available only in LOGLINEAR.

Program 5.1 reads the Banbury data of Table 5.1 and requests a loglinear analysis using the HILOGLINEAR command to fit a model of no interaction.

Program 5.1

```
DATA LIST FREE / Class, Tenure, Vote, Count.
BEGIN DATA.
1 1 1 35   1 2 1 63   1 3 1 22   1 4 1  5
2 1 1 18   2 2 1 29   2 3 1 18   2 4 1 16
3 1 1 12   3 2 1 30   3 3 1 16   3 4 1 24
4 1 1  9   4 2 1 12   4 3 1  8   4 4 1 17
5 1 1  5   5 2 1  1   5 3 1  5   5 4 1  3

1 1 2 5    1 2 2 14   1 3 2  3   1 4 2 10
2 1 2 4    2 2 2  9   2 3 2  4   2 4 2 14
3 1 2 3    3 2 2 44   3 3 2 83   3 4 2 42
4 1 2 7    4 2 2 18   4 3 2 14   4 4 2 45
5 1 2 0    5 2 2  2   5 3 2  5   5 4 2 21
END DATA.
VARIABLE LABELS
  Class 'Occupational Class'.
VALUE LABELS
  Class 1 'I and II' 2 'IIIa, IVa' 3 'IIIb' 4 ' IVb' 5 'V'/
  Tenure 1 'Own outright' 2 'Own on mortgage' 3 'Rented priv'
  4 'Rented council'/
  Vote 1 'Conservative' 2 'Labour'.
WEIGHT BY Count.
HILOGLINEAR Class(1,5), Tenure(1,4), Vote(1,2)
  /DESIGN = Class*Tenure Class*Vote Vote*Tenure.
```

The HILOGLINEAR command starts with a list of the variables to be analyzed, specifying for each the range of values found in the data. For instance, since occupational class has been coded from 1 to 5, the variable list includes Class(1,5). The DESIGN subcommand specifies the model, listing each marginal to be fitted by means of the variable names joined by an asterisk (*). The output includes the observed and the expected (i.e. model) frequencies, which can be checked against Table 5.4.

GLIM

The GLIM program does not use iterative proportional scaling to fit loglinear models, but an alternative and more powerful method. However, the results are the same. Program 5.2 fits the no-interaction model to the data of Table 5.1, and displays the observed and fitted counts. GLIM specifies the model as a set of marginals (made up of the variables concerned, separated by asterisks) connected by plus signs.

Program 5.2

```
$units 40
$data Class Tenure Vote Count
$factor Class 5 Tenure 4 Vote 2
$read
1 1 1 35    1 2 1 63    1 3 1 22    1 4 1  5
2 1 1 18    2 2 1 29    2 3 1 18    2 4 1 16
3 1 1 12    3 2 1 30    3 3 1 16    3 4 1 24
4 1 1  9    4 2 1 12    4 3 1  8    4 4 1 17
5 1 1  5    5 2 1  1    5 3 1  5    5 4 1  3

1 1 2  5    1 2 2 14    1 3 2  3    1 4 2 10
2 1 2  4    2 2 2  9    2 3 2  4    2 4 2 14
3 1 2  3    3 2 2 44    3 3 2 83    3 4 2 42
4 1 2  7    4 2 2 18    4 3 2 14    4 4 2 45
5 1 2  0    5 2 2  2    5 3 2  5    5 4 2 21
$yvar Count
$error p
$fit Class*Tenure + Tenure*Vote + Vote*Class
$display r$
```

The $display directive indicates to GLIM what is to be included in the program output. The parameter "*r*" specifies that the data frequencies, the model frequencies and the residuals (a measure of the difference between the data and model frequencies; see Ch. 7) are to be displayed.

Further reading

The most complete text on loglinear analysis is Bishop et al. (1975); their Chapters 2 and 3 cover the material in the present chapter, but with more mathematics. Knoke & Burke (1980) do the same, but from a more social-scientific standpoint. Davis (1973) works through another example of iterative proportional scaling.

CHAPTER SIX
Choosing and fitting models

Nineteen different models can be fitted to a three-dimensional cross-tabulation. It is clearly necessary to have a way of deciding which of all these models is the most appropriate for a particular data table. A succinct notation to distinguish one model from another is also a necessity. In this chapter, we first look at the range of models that can be fitted to a three-dimensional table and then discuss ways of choosing between them.

In Chapter 5 the construction of two models was demonstrated: a "no-interaction" model and a "no-interaction and no-association between one pair of variables" model. The "no-interaction" model was computed by keeping those three marginal tables that corresponded to each of the three pairs of variables the same in both model and data tables. Symbolically, we can refer to this model as:

$$A*B + B*C + C*A$$

where A, B and C are the three variables. Each of the pairs of variables with asterisks between them corresponds to one of the marginals kept the same in model and data. This notation makes it clear why the "no-interaction" model is also called the "pairwise association" model. Using this model, we are in effect comparing the "real world" with an "imaginary world" in which there may be association between each pair of variables, but no interaction.

The associations between the pairs of variables have been carried over from the "real world" to the "imaginary world" by ensuring that their marginal tables are the same. The marginal tables can be thought of as encapsulating the associations between their variables. Keeping the marginal tables identical in both model and data tables builds these associations into the model. So, to construct a model with only some associations the same as in the data, it is only necessary to arrange for the corresponding marginal tables to be identical in both tables, leaving the remaining marginals unconstrained.

This was illustrated in the second example, in which a model was constructed which carried the associations between class and vote, and sex and vote, over from the "real world", but which had no association between class and sex. The model table was calculated using only two of the three marginal tables, making them identical in both data and model. The remaining marginal, that for class by sex, was allowed to differ from the corresponding data marginal. This model can be referred to symbolically as:

$$A^*B + B^*C$$

if the association between A and C is the one set to zero (for instance, class by sex).

Table 6.1 Models that can be fitted to a data table cross-tabulating variables A, B and C.

A*B*C	Interaction
A*B + B*C + A*C	No interaction, pairwise association
A*B + B*C	No interaction, no association between A and C
A*B + A*C	No interaction, no association between B and C
B*C + A*C	No interaction, no association between A and B
A + B + C	No association

We can now write down in a convenient form some of the models that can be constructed for a three-dimensional table. The six most important are listed in Table 6.1. The interaction model A^*B^*C is the model for an "imaginary world" in which there *is* interaction between A, B and C. The notation implies that the marginal to be kept the same in both model and data is the marginal table relating all three variables. Because this table is the data table itself, this "marginal" table is not really a marginal at all. Since model tables are computed by making the specified marginal table the same as the data, this table is easy to find: it is an exact copy of the data table. It follows that the interaction model for a three-dimensional table *always* fits the data perfectly. (This applies only to three-dimensional tables; if an interaction model is fitted to a data table of more than three variables, it may not fit, depending on the relationships between the variables.) There is, therefore, little point in using the interaction model to examine a three-dimensional table. Similarly, there is little to be gained by comparing a two-dimensional data table against an association model (whose model table is a copy of the data table). That is why we did not bother to fit such a model when investigating in Chapter 4 the relationship between marital status and reported experience of premarital sex. The result of the comparison is a foregone conclusion: it would fit exactly. Models like these which will predictably fit the data are known as *saturated* models.

We have already discussed the pairwise association model and the models that omit one association. The remaining model listed in Table 6.1, $A + B + C$, is the no-association model for a three-dimensional table. This is the model corresponding to an "imaginary world" in which none of the variables is associated with any of the others. Its model table is calculated by requiring that the marginal of each individual variable remains the same in model and data tables, while allowing the marginal tables of A by B, B by C, and C by A to differ. For example, Table 6.2 shows the model table for the Banbury data on class by tenure by vote, using the no-association model

Class + Tenure + Vote

This table shows the frequencies that would be found in an "imaginary world" in which none of the three variables was related to any of the others. Since we have already demonstrated the presence of interaction in these data, it comes as no surprise that this table and the data (Table 5.1) are obviously very different. A little arithmetic would prove that, as required by the model specification, the individual marginals for each of the three variables are the same in this table and the data.

Table 6.2 Model table for model of "no-association" fitted to Banbury data.

Occupational class	Tenure			
	Own outright	Own on mortgage	Rented privately	Rented from council
Voted Conservative				
I and II	11.1	25.1	20.1	22.3
IIIa, IVa	7.91	17.9	14.4	15.9
17.9	17.9	40.6	32.6	36.1
IVb	9.18	20.8	16.7	18.5
V	2.97	6.72	5.39	5.96
Voted Labour				
I and II	11.1	25.0	20.1	22.2
IIIa, IVa	7.89	17.9	14.3	15.9
IIIb	17.9	40.5	32.5	35.9
IVb	9.15	20.7	16.6	18.4
V	2.96	6.70	5.37	5.94

Including a two-variable marginal such as $A*B$ in a model means that the association between these two variables is identical in the model and data tables. Similarly, an interaction can be reproduced by specifying the marginal $A*B*C$ in the model. In much the same way, a single-variable marginal such as A carries over from the data to the model what is called the *main effect* of that variable. The main effect measures the distribution of respondents across the categories of the marginal of a variable. Accordingly, a variable with a marginal in which each cell contains the same value has no main effect.

Main effects, therefore, play a similar rôle for single variables as associations do for pairs of variables, and interactions for three variables. It is rare, however, for a researcher to be directly interested in main effects. Their importance for loglinear analysis lies in the fact that it is usually only models in which all the main effects are included that are examined. To do otherwise would be to test models in which respondents were distributed equally across the categories of one or more variables, an unrealistic assumption for most social science data.

As an example, suppose we wanted to fit a model to the Banbury data in which only class and vote were to be associated, independently of type of tenure. The appropriate model would then be:

Class*Vote + Tenure

the marginal of tenure being included to ensure that its main effect was retained in the model table.

The model $A + B + C$ is often called the "main effects" model instead of the "no-association" model, because it is the model in which only main effects are present.

When we calculated the "no-interaction" model, we ensured that the marginal tables

Class*Tenure + Class*Vote + Tenure*Vote

were the same in model and data tables. In doing this, the marginals of each of the variables class, tenure and vote were forced to be identical in both tables and, thus, the main effects of these variables were carried over into the model. For instance, since the class marginal is just the sum of the cell entries across the categories of vote in the class-by-vote marginal table, making the class-by-vote marginal from the model table equal to the corresponding marginal from the data automatically ensures that the marginal for class is the same in both. The same applies to the marginals of tenure and vote, which are also fixed to be the same as soon as we make the marginal tables including these variables the same.

Generalizing from this example, specifying that a marginal table is to be included in a model has the consequence that the model must also include all those marginals that are themselves marginals to that marginal table. For instance, the model specified by

$$A*B + C$$

sets not only the marginal table *A* by *B* and the marginal for variable *C* but also the marginals of variables *A* and *B*, to be the same in model and data. The model above could in fact be written more fully as:

$$A*B + A + B + C$$

Likewise, the pairwise-association (no-interaction) model can either be written as

$$A*B + B*C + A*C$$

or as

$$A*B + B*C + A*C + A + B + C$$

and these two forms are entirely equivalent.

Table 6.3 Examples of lower-order relatives.

Marginal	Marginal's lower-order relatives
A*B*C	A*B + B*C + A*C + A + B + C
A*B	A + B
B*C	B + C
A*C	A + C
A	none
B	none
C	none

Those marginals that can be derived from a particular marginal table are known as its *lower-order relatives*. Table 6.3 lists the lower-order relatives of each of the marginal tables that can be obtained from a three-dimensional table. Since including a marginal in a model means that all its lower-order relatives are also included, not just any combination of marginals can be used to specify a model. For instance, a model cannot readily be constructed which includes the marginal table *A**B* but does not also include the marginal *A*. This is because *A* is a lower-order relative of *A**B* and is therefore automatically added to the model when *A**B* is specified. In fact, this restriction on the form of models holds only if iterative scaling is used to calculate the model table frequencies. Models in which some lower-order relatives are omitted can be calculated using other more complex methods, but we shall confine discussion in this book to those that do include all their lower-order relatives, the class of models known as *hierarchical* models.

This chapter began with a partial list (Table 6.1) of the models that could be compared with a three-dimensional data table. This list can now be extended by including models with main-effect terms, to give the complete set of 19 hierarchical models (Table 6.4).

Table 6.4 Models for a three-dimensional table, cross-classifying variables *A*, *B* and *C*.

1.	A*B*C
2.	A*B + B*C + A*C
3.	A*B + B*C
4.	A*B + C*A
5.	B*C + C*A
6.	A*B + C
7.	B*C + A
8.	C*A + B
9.	A*B
10.	B*C
11.	C*A
12.	A + B + C
13.	A + B
14.	A + C
15.	B + C
16.	A
17.	B
18.	C
19.	grand mean

Note: The grand mean model, in which no model marginal is constrained, is introduced in Chapter 8.

Choosing a model

With 19 models to choose from when analyzing a three-dimensional table and considerably more for tables of greater dimensionality, it is clear that some way of selecting the most appropriate model is needed. So far, we have chosen models by making assumptions about the likely relationships between the variables concerned. After calculating the model table, we compared it with the data table by making a rough cell-by-cell examination by eye. A more formal method of checking the fit of a model and comparing it with the fit of other models is to use a test of significance.

In classical inferential statistics, tests of significance are used to ascertain whether a statement based on the analysis of a sample (of respondents, for instance) can be generalized to a wider population. However, the logic of these tests makes two strong assumptions: first, that the sample from which data have been obtained is a random sample from a specific population; and secondly, that just one hypothesis, formulated beforehand, is being tested. In most social science research, neither of these assumptions holds good. Even when a sample has been drawn using a random procedure, it is rare for generalizations made from the data to be confined strictly to the population from which the sample has been selected. For instance, the Banbury study used a 6 per cent random sample of Banburians. Nevertheless, the researchers were interested not only in the social structure of Banbury. Their concerns were very much wider, as indicated by their introductory statement that

> an important focus was the fate of the traditional society. It had undoubtedly been under challenge around 1950: had it survived, and in what form? In the "Myth of community studies", Stacey outlined conditions under which a "local social system" might develop and might be destroyed or changed. Banbury in the sixties provided an interesting test case of these ideas. (Stacey et al. 1975: 4)

Moreover, in the course of their discussion, Stacey and her co-authors make frequent comparisons with the results of other community studies, mainly of small towns in the USA. It seems, therefore, that the "population" for their research might reasonably be said to include those resident in urban communities undergoing change from more traditional social structures worldwide, and not just those in Banbury in 1967. These authors are not alone in wishing to make broader generalizations from their data than the sampling procedure they used would seem to warrant. Social science would be the poorer if researchers did feel constrained to restrict their conclusions solely to the samples they had studied.

Social science would also be much hindered if researchers were forced to test one, and only one, hypothesis on the data they had collected. Instead, it is more usual to examine many ideas by testing all of them on the same data. However, the consequence of this more exploratory approach to analysis is that tests of significance lose their original meaning, and the probabilities they generate cannot be relied on as indicators of the generalizability of the hypotheses being tested.

Nevertheless, tests of significance do have an invaluable rôle in loglinear analysis, because they provide the most convenient means of quantifying the comparison of a model table with data. We saw in Chapter 4 that a χ^2 test of significance yields a probability figure that summarizes the cell-by-cell differences between the model and data tables, making allowances (by means of the model's degrees of freedom) for the number of constraints imposed on the fitting of the model. A similar indicator of fit is used to assess loglinear models, although the goodness-of-fit statistic conventionally used is not χ^2, but a related measure called, in full, the log likelihood ratio statistic, or more shortly, *G*-squared (G^2).The log likelihood ratio statistic is defined by:

$$G^2 = 2\sum x_{ij}(\log x_{ij} - \log m_{ij})$$

where, as before, x_{ij} are the data cell frequencies, m_{ij} are the model cell frequencies, "log" represents natural logarithm, and the summation is carried out over all the cells in the table. *G*-squared is generally calculated by the computer program used to perform the analysis. It has a distribution that is almost the same as the χ^2 statistic, and so to look up a *G*-squared probability, a table of the theoretical distribution of χ^2 can be consulted.

Before we can use *G*-squared to examine the fit of models, we need to know how to calculate a model's degrees of freedom. In Chapter 4 it was said that degrees of freedom are an inverse measure of the number of constraints under which the model table frequencies are calculated. The more constraints the model has to satisfy, the lower the number of degrees of freedom. For loglinear models, the constraints are those marginals that are required to be identical in the model and the data. The more marginals specified in a model, the fewer the resulting degrees of freedom.

In fact, a model has degrees of freedom equal to the number of cells in the table minus the total number of degrees of freedom of each of its fitted marginals. Similarly, each marginal table has degrees of freedom equal to its number of cells less the total degrees of freedom of *its* marginals.

To illustrate measuring the fit of a model, let us work through the calculation of the significance figure for the no-interaction model that was developed for the Banbury class-by-tenure-by-vote data (the data table of Table 5.1 and the model table of Table 5.3). Recall that we found this model gave a table that had frequencies similar but not identical to the data table. Because the two tables differed, this added credence to the Banbury study's claim that the data demonstrated statistical interaction between occupational class, type of tenure and voting behaviour. But at the time, we were not able to quantify the difference between the two tables. This we are now ready to do.

G-squared can be calculated (preferably using a computer) by applying the above formula to the model and data table frequencies. It is equal to 41. The calculation of the degrees of freedom is quite straightforward, provided that it is taken step by step. The model table includes five (class) by four (tenure) by two (vote), or 40, cells. The model that was fitted, written out in full to include all the lower-order relatives and abbreviating the variables for the sake of space to *C*, *T* and *V*, is:

$$C{*}T + C{*}V + T{*}V + C + T + V$$

Each term in this specification represents one fitted marginal for which allowance has to be made in calculating the degrees of freedom. There is one further marginal which, in effect, has been fitted when the model table was constructed: the total number of respondents. Since the model table was calculated to include the same number of respondents as the data table, this constitutes a further constraint.

To compute the model's degrees of freedom, the degrees of freedom of each of the fitted marginals must first be found. The degrees of freedom of the number-of-respondents marginal is 1, by definition. The degrees of freedom of the class marginal is equal to the number of cells in this marginal (5) minus the number of degrees of freedom of its marginals. It has only one marginal, the table total with 1 degree of freedom, so that its degrees of freedom is $5-1$, or 4. Similarly, the degrees of freedom of the marginals of tenure and vote are 3 and 1, respectively.

The degrees of freedom of the marginal table, class by tenure, is equal to the number of cells in this table minus the sum of the degrees of freedom of its three marginals: the class marginal, the tenure marginal and the table total. Since the number of cells in the marginal table is 20, its degrees of freedom is:

$$20-(4+3+1) = 12$$

Similarly, the degrees of freedom of the marginal table, class by vote, is 4, and of tenure by vote, 3. The degrees of freedom of the table as a whole can now be calculated as the number of cells overall (40) less the degrees of freedom of C^*T (12), C^*V (4), T^*V (3), C (4), T (3), V (1), and the table total (1), leaving 12 degrees of freedom. Although it is useful to have an idea of the procedure, in practice, the computer program that calculates the model table will also provide the number of degrees of freedom, so you do not normally have to work it out by hand.

Having now both the G-squared value and the degrees of freedom for this model, a table of χ^2 values can be used to look up the corresponding significance level. It is less than 0.01 per cent. A perfectly fitting model would have yielded a significance level of 100 per cent; the no-interaction model has produced such a low significance level that it can be rejected as inappropriate for these data without hesitation.

The reason for a model failing to fit is that one or more relationships that exist in the data have been omitted. However, in this case the only relationship omitted was the interaction between the variables. All the associations have been explicitly included and all the main effects were included as lower-order relatives. Hence it can immediately be concluded that there must be interaction in the data, and that it is this that is causing the lack of fit.

Table 6.5 Attitudes to doctors giving health advice, by gender and social class.

	Males			Females		
Class	I & II	III	IV & V	I & II	III	IV & V
Should give advice*	134	141	40	108	150	55
Only give advice if relevant	27	47	14	47	58	19
Should not give advice	18	36	10	25	46	19

* *Question:* Do you feel doctors should or should not give advice on how to keep healthy (for example, about not smoking, taking exercise and eating the right foods) whenever patients are consulting them about other things?
Source: derived from Arber & Sawyer (1979: Table 9.1).

Table 6.5 relates respondents' attitudes towards doctors giving health advice with their gender and social class (Arber & Sawyer 1979). This table will not be analyzed by first choosing a likely model to examine, as we have done before. Instead, a "brute force" approach will be adopted, calculating in turn the fit of every model that could be applied to the data. This is not an approach that is normally advisable, but it will be instructive to compare the fit of a complete range of models.

The G-squared, degrees of freedom and significance of each model are shown in Table 6.6. As one moves up the table, the models include more and more relationships. The very high G-squared and zero significance levels of the bottom five models (14–18) show, unsurprisingly, that at least one or two main effects are needed for the model to fit the data even approximately. Models 10 and 11 do not fit, because they too omit the main effect of either class or attitude. But, with these exceptions, as further relationships are added the fit improves, the interaction model fitting perfectly.

Table 6.6 Loglinear models fitted to data on attitudes to doctors giving health advice (Table 6.5).

Model†	G^2	d.f.	Significance (%)
(1) A*C*G	0	0	100
(2) A*C + C*G + A*G	4.27	4	37.1
(3) A*C + C*G	10.78	6	9.5
(4) A*C + G*A	7.49	6	27.7
(5) C*G + C*A	10.38	8	23.8
(6) A*C + G	14.43	8	7.1
(7) C*G + A	17.32	10	6.7
(8) G*A + C	14.03	10	17.1
(9) A*C	18.05	9	3.4
(10) C*G	395.11	12	0.0
(11) G*A	187.45	12	0.0
(12) A + C + G	20.99	12	5.1
(13) A + C	24.59	13	2.6
(14) A + G	194.39	14	0.0
(15) C + G	398.76	14	0.0
(16) A	198.01	15	0.0
(17) C	402.38	15	0.0
(18) G	572.18	16	0.0

† Variables abbreviated: A – attitude to doctors giving advice; C – occupational class; G – gender.

Which model should be adopted as the simplest one that fits the data adequately? To decide, a standard of what is to count as an adequate fit is required. The one conventionally used is that models with a significance level of 5 per cent or more are judged to fit well. Using this criterion, models 1–8 fit acceptably and model 12 just scrapes in.

It must be emphasized that the 5 per cent mark is simply a convention. There is no statistical law requiring that a model must achieve a 5 per cent significance level in order to fit the data; the convention is merely a useful way of translating the verbal comment that a model fits reasonably well into a numerical score.

Of the models listed in Table 6.6, the simplest (the one involving the fewest relationships) that passes the 5 per cent mark is model 12, the "no-association" model. But model 8, including one association between attitude and gender, has a much better significance level and is only slightly more complex. We might therefore choose this one to fit the data in preference to the "no-association" model. It is of some interest that neither of these models includes the attitude-by-class relationship that might have been predicted from previous work on doctors and patients.

The effect of the sampling design

G-squared can be used legitimately as a measure of goodness of fit and for model selection only if the data were obtained from a simple random sample. Fortunately, *G*-squared is a fairly robust measure, and most of the random sampling designs likely to be used by social researchers are quite satisfactory bases for loglinear analysis. Two points must be watched, however. First, the sample design itself may fix certain main effects, associations or interactions, and this must be allowed for in the interpretation of loglinear models. Secondly, the formula for *G*-squared may need to be adjusted if the data are collected using a design more complex than a simple random sample.

Sampling designs often require a fixed number of respondents in specified categories. This is usually the case in stratified samples. For instance, the researchers who obtained the data of Table 3.2 on divorce and sexual relations deliberately included roughly equal numbers of divorced and married people in their *sample*. It would obviously not be correct to infer from the near-zero main effect for the marital status variable that there are equal numbers of married and divorced in the *population* at large. Similarly, a stratified sample in which class and gender were the stratifying variables could not be used to obtain results about the relationship between class and gender, because the association between these variables is fixed by design. The usual practice is to include terms representing any relationships that are fixed by design in all the models examined. This ensures that relationships resulting solely from the sample design are taken care of by the model and can thereafter be ignored. Alternatively, the cases in the sample can be *weighted* to reflect the population characteristics (see Clogg & Eliason 1988).

Proportionate stratified samples improve on simple random samples because they ensure that the number of respondents in each stratum is exactly proportional to the number in the population. The equivalent simple random sample will on average be representative of the population, but any one sample may be unrepresentative. The greater likelihood that a stratified sample will be representative is measured by the *design effect* (deff). The effect of stratification is to reduce the deff. Clustering, when only respondents in some randomly selected clusters are sampled, increases the deff. Most large surveys use some combination of stratification and clustering to ensure that the overall deff is not very different from 1.

The formula provided above for G-squared assumes that the data have been obtained from a simple random sample. If the sample has a deff larger than 1, the G-squared given by the standard formula should be reduced by a factor that depends on the size of the design effect; if the deff is less than 1, the G-squared should be increased. The calculation of the appropriate factor is rather complex and depends on the model being tested (see Skinner et al. 1989, Ch. 4). In practice, for most variables, models and most conventional multi-stage cluster sampling designs, such as the UK General Household Survey, the value of G-squared obtained from the standard formula needs to be divided by between 1 and 3 (Holt et al. 1980).

Summary

Loglinear models may be specified in terms of the marginals that are constrained to be equal in model and data tables. For three-dimensional tables, the choice of marginals includes those corresponding to main effects (for individual variables), associations and an interaction. If the latter is specified, the model will necessarily fit exactly. The fit of a model is assessed using a test of significance based on the log likelihood test statistic, or G-squared. The value of this statistic, in conjunction with the number of degrees of freedom of the model, is compared with the χ^2 distribution to establish a significance level. Models with a significance above 5 per cent are generally said to fit the data well.

All the basic elements of loglinear analysis have now been covered. The examples have been confined to three-dimensional tables for simplicity of exposition. However, the analysis of four- and higher-dimensional tables, to be demonstrated in Chapter 8, involves only a natural extension of these ideas and no new principles.

Computer analysis

SPSS

In Chapter 5 there was an example using the HILOGLINEAR command. SPSS also provides the LOGLINEAR command, which offers additional facilities, albeit at the price of some additional complexity in the way analyses are specified and in the computing time taken to calculate the results. Program 6.1 shows how the LOGLINEAR command may be used to obtain the G-squared value for model 8 of Table 6.6, using the data of Table 6.5 on attitudes to doctors giving health advice.

With the LOGLINEAR command the model has to be specified in full, including all the lower-order relatives. For model 8, this means including the marginals for advice and gender as well as the marginal table for advice by gender. The output includes the observed and expected (model) frequencies, with standardized and adjusted residuals (see Ch. 7), and, at the end, two goodness-of-fit statistics, the log likelihood ratio χ^2 (another name for G-squared) and the Pearson χ^2 (the "ordinary" χ^2). The two statistics are almost the same (14.0331 and 14.0317), as is usually the case with large samples.

Program 6.1

```
DATA LIST FREE/ Advice, Class, Gender, Count.
BEGIN DATA.
1 1 1 134
1 2 1 141
1 3 1 40
1 1 2 108
1 2 2 150
1 3 2 55
2 1 1 27
2 2 1 47
2 3 1 14
2 1 2 47
2 2 2 58
2 3 2 19
3 1 1 18
3 2 1 36
```

```
3 3 1 10
3 1 2 25
3 2 2 46
3 3 2 19
END DATA.
VARIABLE LABELS Advice 'Attitude to Doctor giving health
advice'.
VALUE LABELS
  Advice 1 'Should give advice' 2 'Only if relevant'
         3 'Should not give advice'/
  Class 1 'I and II' 2 'III' 3 'IV and V'/
  Gender 1 'Male' 2 'Female'.
WEIGHT BY Count.
LOGLINEAR Advice(1,3), Class(1,3), Gender(1,2)
  /DESIGN=Gender BY Advice, Gender, Advice, Class.
```

GLIM

GLIM displays the G-squared statistic under the name "scaled deviance". Program 6.2 will fit model 8 of Table 6.6. The scaled deviance is displayed as a result of the $fit directive.

In previous examples, the data have been entered as a table in which the levels corresponding to each cell have been included as well as the cell count itself (see e.g. the data for the GLIM example in Chapter 5, or the data in the SPSS program above). It often saves typing if you can get GLIM to calculate and store the levels, so that only the data table cell counts themselves need to be input. This can be done with the %GL (generate levels) function. For example, the directive

```
$calculate Gender = %GL(2,3) $
```

will store the numbers 1 1 1 2 2 2 1 1 1 2 2 2 1 1 1 2 2 2 . . . in the Gender factor, continuing for as many cells as there are in the data (i.e. the number specified in the preceding $units directive). Notice that this %GL function generates the same sequence of 1s and 2s as those that would be typed in for Gender in Program 6.1. The first value in brackets in the %GL function means "generate numbers from 1 to this value", and the second number in brackets means "generate numbers in blocks of this size". Using the %GL function, the analysis for model 8 can be carried out with Program 6.2.

Program 6.2

```
$units 18
$data Count
$read
134 141  40 108 150  55
 27  47  14  47  58  19
 18  36  10  25  46  19
```

```
$factor Advice 3 Class 3 Gender 2
$calculate Advice=%GL(3,6) : Class=%GL(3,1) : Gender=%GL(2,3)
$yvar Count
$error p
$fit Advice*Gender+Class
```

Further reading

Bishop et al. (1975, Ch. 2) give proofs and further results about hierarchical models. Atkins & Jarrett (1979) discuss more fully the limitations of the conventional approach to significance tests. Bishop et al. (1975) also go into further detail about *G*-squared and justify its use in place of χ^2 for loglinear analysis. If many hypotheses are tested on the same data, there is an increased chance that one will be accepted. (With a 5 per cent significance level, it is to be expected that 1 in 20 tests will give the "wrong" result by chance.) This warning applies particularly to selection amongst many competing models, such as those in Table 6.6, and, accordingly, procedures for adjusting significance levels have been developed. A summary can be found in Wrigley (1985: 200–211).

CHAPTER SEVEN

Modelling mobility and change

The previous chapter introduced G-squared as a summary measure of the overall fit of a model table to a data table. It is also instructive to examine the difference between the observed counts in each cell and the corresponding fitted counts. These cell-by-cell differences are known as *residuals*, because they are what remain unaccounted for by the model. This chapter is concerned first with measuring and interpreting residuals and secondly with developing a class of models that fit tables cell by cell on the basis of their residuals. The latter are particularly well suited to the study of mobility and other tables that show how people's characteristics have changed over time.

A residual is the difference between a data value and the value predicted by a model: the difference between the "real" and "imaginary" worlds. At its simplest, the difference can be found by subtracting the data cell count from the corresponding model cell count. However, the simple difference has the disadvantage of not following any sampling distribution and so we cannot tell whether it is larger than might have been expected from the effects of random sampling. A better measure is the ***standardized residual***.

We noted in Chapter 4 that χ^2, one of the statistics that can be used to measure the goodness of fit of a model, has a formula that sums, for all cells in the table, the squared differences between the data and model counts, divided by the model count. The formula quoted in Chapter 4 was

$$\chi^2 = \sum (x_{ij} - m_{ij})^2 / m_{ij}$$

Standardized residuals are the cell-by-cell components that are squared and summed when calculating this χ^2 statistic. So the residual for cell *i,j* is:

$$z_{ij} = (x_{ij} - m_{ij}) / \sqrt{m_{ij}}$$

These residuals have an approximately normal distribution with a mean of 0 and a standard deviation of 1, so, as a rule of thumb, those greater than 2 or less than −2 are larger than would be expected if the model fitted the data.

Table 7.1 Standardized cell residuals obtained by fitting a model of "no interaction" to the Banbury data of Table 5.1.

	Tenure			
Occupational class	Own, outright	Own, on mortgage	Rented privately	Rented from council
Voted Conservative				
I and II	−0.22	0.17	1.11	-1.50
IIIa, IVa	-0.36	0.17	1.11	-1.50
IIIb	0.86	-0.03	-2.19	2.21
IVb	-0.46	0.28	0.51	0.29
V	0.94	0.28	1.08	-1.31
Voted Labour				
I and II	0.69	−0.34	−1.67	2.01
III, IVa	1.02	0.06	−1.32	0.55
IIIb	−1.11	−0.03	1.36	−1.22
IVb	0.61	0.25	−0.34	−0.17
V	−1.30	0.25	−0.72	0.78

Table 7.1 shows the standardized residuals obtained from fitting a model of no interaction to the Banbury data introduced in Chapter 5. This model did not quite fit, so it would be interesting to see where the lack of fit is greatest. There are three large residuals, of over 2 in absolute magnitude, each indicating a cell for which the interaction effect is particularly marked. There are more skilled-manual (class IIIb) Conservative-voting council tenants and more professional and managerial (classes I and II) Labour-voting council tenants in the data table than would be expected if there were no interaction. It is among these categories of people that the joint effect of class and tenure has the most influence on voting behaviour.

In addition to examining individual residuals, it is often helpful to plot them. One revealing way to do this is to use a "three-dimensional" histogram in which the magnitude of each residual is shown by the height of a column, as in Figure 7.1. This technique shows clearly the parts of the table that have particularly large residuals, but it can be used effectively only for two- and three-dimensional tables. For more

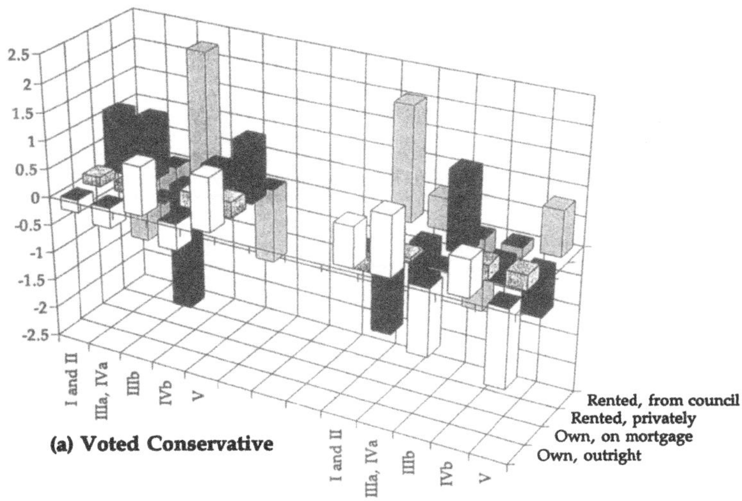

Figure 7.1 Plot of residuals from fitting a no-interaction model to the Banbury data.

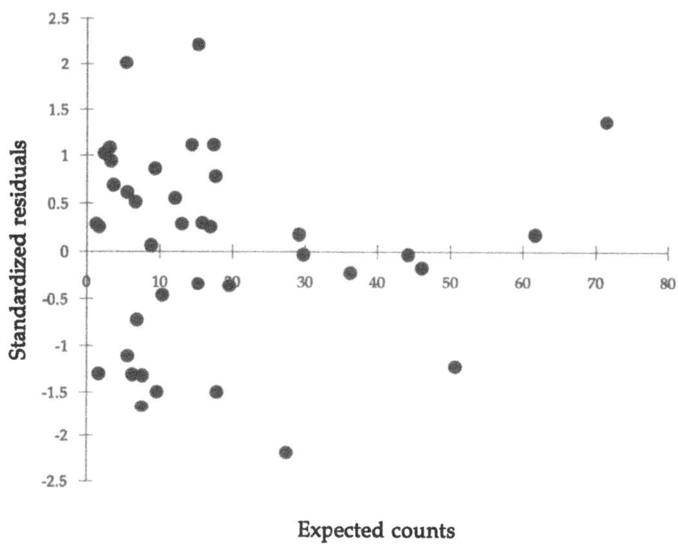

Figure 7.2 Plot of residuals against expected counts from fitting a model of no interaction to the Banbury data.

complex tables, the residuals can be plotted against the expected counts (Fig. 7.2). If the model is appropriate for the data, the plot should show the points evenly scattered without any clear pattern. (For example, the points should not be clustered around a line.)

Standardized residuals are only approximately normally distributed, and the rule of thumb that those with an absolute value larger than 2 are "large" is only a rough guide. Some cells that do not fit well may not show up using this criterion. A somewhat better measure is the *adjusted residual*, which is obtained by dividing the standardized residual by its estimated standard error, the latter being a measure of the variability of the residual among different samples from the same population (see Ch. 9). The adjusted residual is also normally distributed with unit variance and a zero mean, so that an absolute value greater than 2 indicates a cell that does not fit.

Structural values

Data tables sometimes include cells whose frequencies are known before any data are collected. For instance, in a cross-classification of criminals by the crime for which they have been convicted and their gender, we know immediately that the cell corresponding to "rape by females" will be empty. Likewise, in a table of delinquency by age, the cell corresponding to delinquents under 11 years old will be empty since young children are not legally responsible for their "criminal" acts. Such pre-specifiable cell values are known as *structural values* (or "structural zeros", since the fixed value is usually zero). By no means are all zero cells in a data table structural zeros, but only those in which a zero frequency is a logical necessity. Cells in which no observations have been recorded because the sample happened not to include members with that combination of categories are known as *sampling zeros*.

Distinguishing structural from sampling values is important because structural values place constraints on the model table. For instance, a model table that included a number of female rapists would not be satisfactory for the crime/gender data. The model table must be constrained so that it has the appropriate frequencies in the structural cells. Fortunately, this is fairly easy to achieve. The structural cells are just avoided when fitting the model. The calculation of model table frequencies proceeds as though the structural cells were absent from the table;

their values have no influence at all on the calculation of the other frequencies.

The presence of structural cells does, however, affect the degrees of freedom of the model. The degrees of freedom are reduced by one as compared with the normal number for every structural value in the data table, provided that no marginal table includes more than one count obtained by summing over structural values. If this condition is not met, the adjustments to the degrees of freedom that are needed are less straightforward to calculate. Bishop et al. (1975) discuss these awkward cases in detail.

Partitioning and topological models

An extension of the idea of structural values leads to a technique for analyzing a data table with the help of models in which the data are *partitioned* into two or more sections, each of which is fitted separately. These models are often called *topological models* (Erikson & Goldthorpe 1978; Hauser 1978).

Table 7.2 shows a cross-tabulation of data from a survey in 1972 of men aged 20–64 in England and Wales (Goldthorpe 1980). The column variable is the respondent's own social class and the row variable is the class of the respondent's father when the respondent was aged 14. Social class was assigned on the basis of occupation. Class I includes higher-grade professionals, administrators, managers in large enterprises, and large proprietors. Class II consists of lower-grade professionals, administrators and technicians, and managers in small business. Class III is routine non-manual employees: the bulk of white-collar workers. Class IV comprises small proprietors, including farmers and self-employed artisans. Class V consists of supervisors of manual workers and class VI, skilled manual workers including those who have served apprenticeships and have other kinds of training. Class VII covers all manual workers in semi- and unskilled grades and agricultural workers. The table is an example of a common kind of data in which the same concept is measured at two points in time and then cross-tabulated.

Although the table is interesting for the light it can shed on the openness or otherwise of the class structure, there is little we can do to explore it using the techniques discussed so far. We can notice that the distributions of people in the two marginals are different, reflecting

Table 7.2 Intergenerational class mobility: England and Wales, 1972.

	Son's (i.e. respondent's) class							
Father's class	I	II	III	IV	V	VI	VII	Total
I	311	130	79	53	33	37	45	688
II	161	128	66	39	53	59	48	554
III	128	109	89	54	89	108	117	694
IV	167	151	106	324	116	192	273	1329
V	154	147	109	83	170	229	190	1082
VI	202	228	216	170	319	788	671	2594
VII	162	194	205	164	311	587	870	2493
Total	1285	1087	870	887	1091	2000	2214	9434

Source: derived from Goldthorpe (1980, Table 4.2).

changes in the occupational structure between the time when the respondents were 14 and the time of the survey. We can test for the main effects of the variables (both highly significant) and for association between the variables (also highly significant). The main effects just indicate that the classes are not equal in size. The no-association model does not fit this table nor most social mobility tables because it leaves large residuals along the diagonal from top left to bottom right. This is a sign of what Goodman (1965) calls "status persistence" or "status inheritance", the tendency for children to remain in the same class as their parents. These results are to be expected from the nature of the variables: it is not surprising that respondent's class is associated with father's class. What we need is a way of looking at the internal structure of the table, to see in more detail how fathers' and sons' classes are related.

A clue about this internal structure can be seen in Table 7.3, the standardized residuals obtained from fitting a model of no association. As expected, there are large positive residuals in the cells lying along the *leading diagonal* of the table, i.e. the cells for the respondents who are in the same class as their fathers were. It may be that the association is accounted for by the tendency for sons to stay in their fathers' class. Once this is put aside, there may be relatively little relationship between class of origin and class of destination. This possibility can be investigated with a second model which differs from the first in that the leading diagonal cells are defined to be structural. This device causes the diagonal cells to be omitted from the fitting, so allowing the association in the off-diagonal cells to be examined alone (see Table 7.4).

Table 7.3 Standardized residuals from fitting a model of simple independence to the data of Table 7.2.

No-association model: $G^2 = 1744$; d.f. $= 36$; significance: 0%

	Son's (i.e. respondent's) class						
Father's class	I	II	III	IV	V	VI	VII
I	22.4	5.7	2.0	−1.5	−5.2	−9.0	−9.1
II	9.8	8.0	2.1	−1.8	−1.3	−5.4	−7.2
III	3.4	3.2	3.1	−1.3	1.0	−3.2	−3.6
IV	−1.0	−0.2	−1.5	17.8	−3.0	−5.3	−2.2
V	0.5	2.0	0.9	−1.9	4.0	−0.0	−4.0
VI	−8.1	−4.1	−1.5	−4.7	1.1	10.2	2.5
VII	−9.6	−5.5	−1.6	−4.6	1.3	2.5	11.8

Table 7.4 Standardized residuals from fitting a model of quasi-independence.

No-association model, leading diagonal not fitted: $G^2 = 661$; d.f. $= 29$; significance: 0%

	Son's (i.e. respondent's) class						
Father's class	I	II	III	IV	V	VI	VII
I	0	10.3	5.0	3.1	−3.1	−5.6	−5.2
II	12.6	0	2.1	−0.0	−1.3	−4.2	−5.8
III	4.6	2.6	0	−0.0	0.1	−2.8	−2.8
IV	2.4	1.1	−0.8	0	−2.2	−2.3	1.9
V	1.8	1.3	−0.1	−0.1	0	0.7	−2.8
VI	−5.1	−3.5	−1.5	−1.0	1.2	0	7.2
VII	−6.3	−4.3	−1.0	−0.3	2.3	7.3	0

The result of fitting this model (known as a "quasi-independence" model) is disappointing: it does not fit much better than the simple independence model. We can conclude that, even when sons move out of their fathers' class, their class of origin affects which class they move into. Incidentally, the degrees of freedom of this model are seven fewer than for the independence model because there are seven fewer cells to fit.

The same technique can be used to fit models in which cells other than those on the diagonal are to be omitted. For example, it is possible to fit just the triangle of cells under the diagonal, ignoring the rest of the table, if the issue of interest is whether the class of the sons is independent of the class of the fathers, among those who were downwardly mobile. Fitting the upper triangle would test whether destination is independent of origin among those who were mobile up the class schema. In all such models, the hypothesis is that there is no association between the row and column variables within the cells that are fitted. Tables like these in which some cells are empty or ignored are known as "incomplete tables".

In fitting the quasi-independence model, we ignored the cells on the diagonal by treating them as structural. Instead of ignoring them, we could have treated each cell on the diagonal separately, ensuring a perfect fit for these cells. The results would have been the same. To understand how this works, imagine expanding the two-dimensional social mobility table by adding seven empty tables behind the original table, thus changing the table into a stack of eight tables, one behind the other. Then distribute the cell counts from the original table so that all the counts from the off-diagonal cells remain in the front slice of the stack, and a single count from each cell on the diagonal is moved from the front table into one of the subsequent seven slices of the stack. If all the empty cells in the stack are treated as structural, the first slice is identical to the table we modelled above (cells counts in all off-diagonal cells; structural cells along the diagonal). The remaining slices of the stack each have only one cell count in them - all the other cells in the slice are structural. These slices are often called *partitions*.

We now have a three-dimensional table with a lot of structural zeroes. If the new third dimension is given a name, say, Q, it can be treated as though it were a variable when fitting models. The quasi-independence model becomes:

Father's Class + Son's Class + Q

The partitioning variable, Q, has an interpretation: it represents the degree to which father's class is inherited by sons. Each partition of the stack includes those respondents for which the value of Q is the same; in other words, these respondents all inherit their fathers' class to the same degree.

The distribution of cells to partitions can be shown in a table where the numbers in the cells indicate the partitions into which the cell counts have been placed. This table is known as a *levels matrix*. For the quasi-independence model, the matrix is shown in Table 7.5.

Other patterns of levels can also be fitted. For example, classes I and II both include professionals and administrators, and we might hypothesize that there will be an extra tendency for those with fathers in one of these two top classes to remain there, because of their advantages of income and education. In other words, we might suppose that

Table 7.5 Levels matrix for model 2: quasi-independence.

	Son's (i.e. respondent's) class						
Father's class	I	II	III	IV	V	VI	VII
I	2	1	1	1	1	1	1
II	1	3	1	1	1	1	1
III	1	1	4	1	1	1	1
IV	1	1	1	5	1	1	1
V	1	1	1	1	6	1	1
VI	1	1	1	1	1	7	1
VII	1	1	1	1	1	1	8

there is a barrier keeping sons with class I and II fathers from leaving these classes and preventing sons with fathers in other classes from entering. Noting also that classes VI and VII are manual workers, who often lack educational or financial resources to move up the class structure, there might be an extra tendency for sons with fathers in these classes to remain there. For simplicity, we shall suppose that elsewhere in the table there is a tendency for sons to remain in their fathers' class, but that if they do change class, their destination is independent of their fathers' class.

These hypotheses can be examined with a model that involves the pattern of levels shown in Table 7.6. This model fits better than the previous one (G^2 = 351 with 31 degrees of freedom), although it is still not satisfactory. Clearly, the assumption is incorrect that son's class is independent of father's class in the areas assigned to level 1. Progress can be made by moving these off-diagonal cells to other levels. Goldthorpe & Payne offer a carefully argued proposal for the levels matrix of Table 7.7 (Goldthorpe 1980). This gives a G-squared of 31.4 with 29 degrees of freedom and fits well. The fact that this model fits means that there is independence within each set of cells assigned to the same level in the levels matrix (i.e. assigned to the same partition of the three-dimensional table). Within these cells, there is a constant amount of inheritance of father's class position, or *density* (Hauser 1979). The meaning of density can be understood by comparing how cell counts

Table 7.6 Levels matrix for model 3.

	Son's (i.e. respondent's) class						
Father's class	I	II	III	IV	V	VI	VII
I	2	2	1	1	1	1	1
II	2	2	1	1	1	1	1
III	1	1	3	1	1	1	1
IV	1	1	1	4	1	1	1
V	1	1	1	1	5	1	1
VI	1	1	1	1	1	6	6
VII	1	1	1	1	1	6	6

Table 7.7 Levels matrix for Goldthorpe's eight-level design.

	Son's (i.e. respondent's) class						
Father's class	I	II	III	IV	V	VI	VII
I	1	3	4	5	7	8	8
II	3	3	4	6	6	7	7
III	4	4	4	6	5	6	6
IV	5	5	5	2	6	6	6
V	5	4	4	6	4	5	6
VI	7	6	5	6	5	4	5
VII	7	6	5	6	5	5	4

Source: Goldthorpe (1980, Fig. 4.1).

are modelled in a partitioned table and in a simple independence model. The expected count in a cell for a simple independence model is made up of two components: one dependent on the class of origin (which is the same for all destination classes) and one dependent on the class of destination (which is the same for all origin classes). For a partitioned model, the expected count also depends on the density of the partition to which the cell has been assigned. The value of the density can be calculated using the techniques for estimating parameters to be described in Chapter 9. Goldthorpe chose the numbering scheme for his levels such that level 1 has the highest density (highest tendency for sons to inherit their fathers' class) and level 8 the lowest.

Thus, the cell at the top left of the mobility table, for Class I fathers and Class I sons, is where there is most "status persistence", or least mobility. As expected, the lower levels (least mobility) are in cells along the diagonal and the higher-levels are in cells at the top right and bottom left.

The model above fitted one additional partitioning variable, Q. Still more complex models can be constructed which add further partitioning variables to the model. Each variable is defined by a separate levels matrix designed to capture a specific effect on mobility rates. Some care is needed in constructing these models, however. First, it is easy for the model to become so complicated that it is very difficult to interpret. Secondly, each new level takes away a degree of freedom, and it is possible to run out of degrees of freedom. (A model with zero degrees of freedom always fits.) And thirdly, the flexibility of these models, arising from the fact that each cell can be allocated to any level, means that it is possible to find patterns of levels that fit the data, but are not interpretable sociologically. It is therefore especially important to develop a sociologically grounded hypothesis about mobility and to translate this into levels matrices before testing it, rather than to create matrices in an *ad hoc* fashion from inspection of the data.

In a later study of the same mobility data, Erikson & Goldthorpe (1978) use a model for which they define eight levels matrices, each specified in terms of two levels, to represent different mobility effects (Table 7.8). Two of these matrices indicate the relative desirability of classes of destination and the relative advantages of classes of origin. Three matrices account for inheritance effects, which increase the likelihood that individuals will be found in the class in which they originated. One matrix represents a barrier between agricultural and other occupations. And the final two matrices capture a number of processes that give rise to affinities or barriers between specific classes. Erikson & Goldthorpe hypothesize a barrier between agricultural workers and the professional and administrative classes, and affinities between the professional and administrative classes on the one hand, and routine non-manual employees on the other, and between the classes of skilled, semi-skilled and unskilled industrial workers. The resulting model does not fit according to conventional criteria (the G-squared is 68.3 with 28 degrees of freedom), but the sample size is large (9,434) and the residuals do not appear to have any important structure (see Table 7.9).

Table 7.8 Arrangement of hierarchy, inheritance, sector and affinity effects for modelling intergenerational mobility.*

	Son's class						
Father's class	I + II	III	IVa+b	IVc	V + VI	VIIa	VIIb
I+II	IN1, IN2	HI1, AF2	HI1, AF2	HI1, SE	HI1	HI1, HI2	HI1, HI2, SE, AF1
III	HI1, AF2	IN1		SE		HI1	HI1, SE
IVa + b	HI1, AF2		IN1, IN2	SE, AF2		HI1	HI1,SE
IVc	HI1, HI2, SE	HI1, SE	HI1, SE, AF2	HI1, IN1, IN2, IN3	HI1, SE	SE, AF2	
V + VI	HI1			SE	IN1	HI1, AF2	HI1, SE
VIIa	HI1, HI2	HI1	HI1	HI1, SE	HI1, AF2	IN1	SE
VIIb	HI1, HI2, SE, AF1	HI1, SE	HI1, SE	HI1	HI1, SE	SE, AF2	IN1

* Codes show the cells in which the indicated matrix is at the second level.
Source: Erikson & Goldthorpe (1992, Table 4.3).

Table 7.9 Observed and fitted counts for intergenerational mobility in England and Wales.*

	Son's class						
Father's class	I+II	III	IVa+b	IVc	V+VI	VIIa	VIIb
I+II	730 (731.2)	145 (113.7)	79 (82.2)	13 (3.1)	182 (208.9)	91 (100.8)	2 (2.1)
III	237 (229.3)	89 (107.4)	51 (48.4)	3 (2.9)	197 (189.9)	113 (111.1)	4 (5.0)
IVa + b	252 (270.7)	81 (80.0)	187 (185.8)	8 (5.2)	222 (224.3)	144 (131.2)	8 (5.9)
IVc	66 (52.1)	25 (28.5)	30 (31.9)	99 (99.0)	86 (80.9)	81 (100.9)	40 (33.7)
V + VI	731 (728.6)	325 (328.5)	248 (237.7)	5 (14.0)	1506 (1499.0)	839 (843.9)	22 (24.4)
VIIa	330 (334.1)	187 (182.7)	122 (132.2)	9 (7.8)	802 (802.3)	685 (672.2)	15 (18.7)
VIIb	26 (25.9)	18 (30.2)	23 (21.9)	10 (14.9)	96 (85.8)	114 (106.9)	56 (57.4)

* Fitted values in brackets.
Source: Robert Erikson, unpublished data derived from the Oxford National Occupational Mobility Inquiry (Goldthorpe 1980).

Stier & Grusky (1990) describe an even more complex model to fit a social mobility table that cross-tabulates the first jobs with the current jobs in 1973 of nearly 18,000 people drawn from the US male economically active population. Occupations were divided into 18 categories, giving a 18 × 18 cell mobility table of first occupation by current occupation. This table was then modelled using levels matrices representing persistence in various occupational and sectoral groupings.

The first set of matrices represented "occupational persistence". This is the tendency for people to remain within a broad occupational grouping although not necessarily within the same firm or industry, a tendency which Stier & Grusky suggest is enhanced in jobs with general skills, especially when these are validated by professional associations. The second set of matrices relates to "sub-occupational persistence", the likelihood that people will remain within a detailed occupational category because of their investment of capital (the case with farmers and owners of businesses) or education and training (the case for those with rather specialized jobs). The third set concerns persistence within the manual and non-manual strata, and the fourth covers persistence within finer divisions of this dichotomy: professionals and managers, sales and clerical, crafts, service workers and labourers, and farmers. The fifth set of matrices allows for persistence within the "core" and "periphery" industrial sectors, and the final set caters for a mixed bag of affinities between occupational categories resulting from a variety of other features of the occupational order, for example, affinities between managers and professionals, craft workers and professionals, and farm labourers and non-manual occupations. The resulting model fits the data well, with a G-squared of 620 and 251 degrees of freedom. The authors argue that their matrices capture the effects of various specialized resources such as vocational training and occupational credentials on social mobility and persistence, and give an economical and interpretable picture of mobility patterns within occupational careers.

Although the examples in this chapter were concerned with social mobility, there is no reason for these techniques to be confined to this kind of data. They can be applied to tables that are not "square" and to tables including more than two variables. Indeed, topological modelling may be appropriate and useful whenever there is reason to believe that certain combinations of categories of variables have consequences different in kind from those of other combinations.

Summary

Whenever a model is fitted to data, it is important to look at the residuals, the differences between the data and the model table counts. Examination of residuals can reveal where models are fitting badly and can identify particular cells that need special treatment. Tables with "special" cells may be analyzed by preserving those cells from modification when fitting a model. This technique can be used with advantage in dealing with structural cells, that is, cells in which the frequency is fixed by design.

The idea of deleting certain cells from the fitting of a model can also be used to examine incomplete tables and tables where it is hypothesized that different effects operate in different sections of the table. In general, a table can be divided into two or more partitions, as defined by a levels matrix, and each partition fitted separately. Alternatively, a partitioning variable can be used to specify the level of each cell in a table and this variable can be added to the model. More complex designs in which there are several levels matrices are also possible. These topological models are particularly powerful for the analysis of mobility, but may also be useful in other contexts.

Computer analysis

The programs that follow illustrate the use of topological models with the social mobility data of Table 7.2. First, a simple independence model is fitted and then, by defining the leading diagonal cells as structural, a quasi-independence model. SPSS version 4 is not able to fit models using the levels matrix approach. Using GLIM, a second quasi-independence model is constructed based on a levels matrix. Goldthorpe's eight-level matrix (Table 7.7) is fitted to the data. Program 7.3 replicates Erikson & Golthorpe's analysis with the set of levels matrices shown in Table 7.8.

SPSS

The DATA LIST command reads the mobility data. A no-association model is fitted to the data table with the first LOGLINEAR command.

In order to declare the leading diagonal cells to be structural for the quasi-independence model, a weighting variable has to be constructed.

In Program 7.1, the COMPUTE and IF commands create a variable, called OffDiag, which is equal to 1 for every cell unless the cell is on the diagonal; i.e., the cell has equal Father and Son category values when it is set to 0. This weighting variable is used in the second LOGLINEAR command, which fits an independence model to those cells for which OffDiag is equal to 1.

Program 7.1

```
* Social mobility data from Goldthorpe (1980).
DATA LIST FREE/ Father Son Count.
BEGIN DATA.
1 1 311
1 2 130
1 3  79
[data omitted: see Table 7.2]
7 4 164
7 5 311
7 6 587
7 7 870
END DATA.
WEIGHT BY Count.
* Simple independence model.
LOGLINEAR Father, Son(1,7)
 /DESIGN= Father, Son.
* Quasi-independence model, defining leading diagonal
   as structural.
COMPUTE OffDiag = 1.
IF (Father = Son) OffDiag = 0.
LOGLINEAR Father, Son(1,7)
 /CWEIGHT= OffDiag
 /DESIGN= Father, Son.
```

GLIM

Program 7.2 begins by reading in both the data table and the Goldthorpe levels matrix. The line beginning $c is a comment for people reading the program and is ignored by GLIM. The first $fit directive tests a main-effects-only (no-association) model on the mobility data. The following $calculate directive sets the variable Weight to 1 for cells in which the Father category is not equal to the Son category and to 0 for other cells. (GLIM uses the symbol "\=" to mean "not equal to".) The data are then weighted ($weight) with this variable, which has the effect of omitting cells on the leading diagonal from the fitting because the weights for these cells are 0. The quasi-independence model is then fitted.

The second $weight directive cancels the effect of the first one. The

$calculate directive computes a levels matrix and puts its values into the variable *Q*. The calculation assigns 1 plus the value of the Father variable to those cells that have father's and son's class the same, and 1 to all other cells. This results in a levels matrix appropriate for a quasi-independence model (see Table 7.5). The $tprint directive displays the matrix. A model that includes the *Q* variable is fitted. The model table is identical to the one calculated using zero weights along the diagonal.

Finally, the Goldthorpe levels matrix which was read into the Level factor is displayed and fitted to the data.

Program 7.2

```
$c Social mobility data from Goldthorpe (1980), Table 4.2
$units 49
$c Level indicates assignment of cells to levels from
$c    Goldthorpe (1980), Figure 4.1
$data Father Son Level Count
$factor Father 7 Son 7 Level 8
$read
1 1 1 311
1 2 3 130
1 3 4  79
[ data omitted: see Tables 7.2 and 7.7]
7 4 6 164
7 5 5 311
7 6 5 587
7 7 4 870
$yvar Count
$error p
$c Simple independence model
$fit Father + Son
$display r
$c Quasi-independence model, defining leading diagonal
$c  as structural
$calculate Weight = Father /= Son
$weight Weight
$fit Father + Son
$display r
$weight
$c Quasi independence model using levels matrix
$calculate Q = %if(%eq(Father, Son), Father, 0) +1
$factor Q 8$
$tprint Q Father ; Son
$fit Father + Son + Q
$display r
$c Goldthorpe's (1980) eight level model
$tprint Level Father ; Son
$fit Father + Son + Level
$display r
```

Program 7.3 fits Erikson & Goldthorpe's (1992) core social fluidity model, shown in Table 7.8, to the data of Table 7.9. Each of the eight levels matrices are read in, one after the other, and then fitted to the data.

Program 7.3

```
$c Erikson & Goldthorpe's (1992) core fluidity model
$units 49
$factor Father 7 Son 7
$calc Father = %GL(7,7)  : Son = %GL(7,1)
$data Count
$read
730  145   79   13  182   91    2
237   89   51    3  197  113    4
252   81  187    8  222  144    8
 66   25   30   99   86   81   40
731  325  248    5 1506  839   22
330  187  122    9  802  685   15
 26   18   23   10   96  114   56
$factor HI1 2 HI2 2 IN1 2 IN2 2 IN3 2 SE 2 AF1 2 AF2 2 $
$data HI1
$read
1 2 2 2 2 2 2
2 1 1 1 1 2 2
2 1 1 1 1 2 2
2 2 2 2 2 1 1
2 1 1 1 1 2 2
2 2 2 2 2 1 1
2 2 2 2 2 1 1
$data HI2
$read
1 1 1 1 1 2 2
1 1 1 1 1 1 1
1 1 1 1 1 1 1
2 1 1 1 1 1 1
1 1 1 1 1 1 1
2 1 1 1 1 1 1
2 1 1 1 1 1 1
$data IN1
$read
2 1 1 1 1 1 1
1 2 1 1 1 1 1
1 1 2 1 1 1 1
1 1 1 2 1 1 1
1 1 1 1 2 1 1
1 1 1 1 1 2 1
1 1 1 1 1 1 2
$data IN2
$read
2 1 1 1 1 1 1
1 1 1 1 1 1 1
```

```
1 1 2 1 1 1 1
1 1 1 2 1 1 1
1 1 1 1 1 1 1
1 1 1 1 1 1 1
1 1 1 1 1 1 1
$data IN3
$read
1 1 1 1 1 1 1
1 1 1 1 1 1 1
1 1 1 1 1 1 1
1 1 1 2 1 1 1
1 1 1 1 1 1 1
1 1 1 1 1 1 1
1 1 1 1 1 1 1
$data SE
$read
1 1 1 2 1 1 2
1 1 1 2 1 1 2
1 1 1 2 1 1 2
2 2 2 1 2 2 1
1 1 1 2 1 1 2
1 1 1 2 1 1 2
2 2 2 1 2 2 1
$data AF1
$read
1 1 1 1 1 1 2
1 1 1 1 1 1 1
1 1 1 1 1 1 1
1 1 1 1 1 1 1
1 1 1 1 1 1 1
1 1 1 1 1 1 1
2 1 1 1 1 1 1
$data AF2
$read
1 2 2 1 1 1 1
2 1 1 1 1 1 1
2 1 1 2 1 1 1
1 1 2 1 1 2 1
1 1 1 1 1 2 1
1 1 1 1 2 1 1
1 1 1 1 1 2 1
$yvar Count
$error p
$fit Father + Son + HI1 + HI2 + IN1 +IN2 + IN3 + SE + AF1 +AF2
$disp e
$tprint Count ; %fv Father ; Son
```

Further reading

Hauser (1979) provides a brief introduction to the use of partitioned models to investigate social mobility. Hout (1983) considers modelling mobility tables in greater detail, with an emphasis on quasi-independence models. Bishop et al. (1975, Ch. 5) give many examples of the analysis of incomplete tables. Duncan (1965) shows how partitioning may be used to study the effect of individual cells of a table.

CHAPTER EIGHT
High dimension tables

In previous chapters, we have looked at a variety of two- and three-dimensional tables and have shown how models can be developed to fit them. The techniques can be readily extended to tables including more variables. When analyzing such complex tables, it is especially important to work through the range of appropriate models in a systematic way. A technique known as *forward selection* can be used to guide the choice of the best model, and this chapter will demonstrate forward selection on a four- and a five-dimensional table. The first example will also serve as a good illustration of the value of loglinear analysis for exploring tables involving more variables than can conveniently be handled by other methods, such as elaboration.

We shall look, first, at some of the factors that might influence people's ideas about how satisfactory their standard of living and financial circumstances are. Obviously, one of the most important of these is likely to be their income. But reference group theory (Merton 1968, Runciman 1966) suggests that people assess their standard of living not only in absolute terms, but also by comparing their situation with that of others. Hence the breadth of people's horizons and the choices they make about whom they compare themselves with are also relevant. Furthermore, people's past experiences - in particular, their parents' standard of living and changes in their own standard of living - may well be important. So, too, could be their level of education. All these factors could be expected to influence judgements about how satisfied people are with their standard of living.

These possibilities can be explored using data from the National Opinion Research Centre's General Social Survey. NORC (ICPSR 1978) annually questions a large random sample of US residents on a wide range of issues and makes the resulting data publicly available for further secondary analysis (Dale et al. 1988).

The questions include five relevant to our interests. Respondents were asked how satisfied they were about their current financial situation, and also how they thought their income compared with that of other US families. They were questioned about how their family income had compared with that of other families when they were aged 16, and whether they thought their own family circumstances had been getting better or worse during the last few years. These four questions together provide data on respondents' perceptions of their financial circumstances at present and in the past, relative to the perceived standard of living of their compatriots. We shall also use data from a question that inquired about the respondents' level of education, as measured by the number of years of school or college they had completed.

The responses from six similar surveys conducted in 1972–8 were combined, giving a total sample of 10,307 randomly selected US citizens aged between 18 and 99. The data were cross-classified to yield a five-dimensional table of four variables concerned with perceptions of family circumstances and one variable indicating educational level. This data table includes 300 cells and is too large to reproduce here. It is the basis for the exploration of the factors affecting satisfaction with current standard of living that follows.

The prime concern will be with the interrelationships between the five variables just described. Of these variables, the first four are all concerned with respondents' *perceptions* of their relative income as compared with others. We shall not here be analyzing the effect of their *actual* income. There is undoubtedly an association between respondents' income and their satisfaction with their financial circumstances. Cramer's V, a measure of strength of relationship appropriate for tables with more than two levels per variable (see Ch. 4), is equal to 0.21 for the association between income and satisfaction with current income, indicating a moderate association. However, this is incidental to the focus of this example, which is concerned more with the relationships between various perceptions and changes in perceptions.

Analyzing a five-dimensional table is much the same as analyzing smaller ones. One difference lies in the range of components that are available to build models. With the three-dimensional tables encountered so far, there has been only one possible interaction effect, the one representing the interaction between the three variables. In five-dimensional tables there can be interactions between any combination of three of the five variables. There can, therefore, be ten different interactions. Moreover, with five variables, higher-order effects are also possible. Thus, there may be effects (known as *second-order interactions*) representing the difference in the interaction between three variables for the various levels of a fourth variable. (Compare the definition of an interaction as the difference in association between two variables for various levels of a third.) Finally, a third-order interaction may exist when a second-order interaction differs according to the levels of a fifth variable. Fortunately, since they are often rather difficult to interpret in readily comprehensible sociological terms, second- and third-order interactions are rarely found in practice.

Returning to the example, although reference group theory suggests that there are quite likely to be associations between most of the five variables and that there may even be interactions among them, it is not clear which particular model should be choosen for examination first. Theoretical preconceptions are not sufficiently precise to justify the selection of one specific set of interrelationships as an initial model. This is frequently the case when dealing with real research problems. Because we do not know much about the likely structure in the data, we shall use a model selection procedure known as *forward selection*.

Forward selection

Forward selection involves first examining the very simplest model for its fit, and then successively adding further effects until sufficient have been included for the fit to be good. The simplest model that can be applied to a data table is the *grand mean* model. This is a model in which all the cell frequencies are exactly the same so that respondents are distributed uniformly through the table. Such a model is highly unlikely to be a good fit to any data encountered in practice. The grand mean model is therefore rarely proposed for real data. Nevertheless, because it is always the worst-fitting of any model, it does serve a useful purpose as a point of comparison or benchmark against which more complex models may be assessed. The grand mean model table is straightforward to calculate: the total number of respondents is divided by the number of cells and the resulting frequency is put into each cell of the table.

The first step of forward selection is to fit the grand mean model (model 1 in Table 8.1). Having found, without surprise, that it does not fit, we try a main-effects model in which each of the single-variable marginals is included (model 2). This model does not fit well either, so we move to a pairwise-association model (model 3) and then to a model including all possible interactions (model 4). At last we have a model that fits well, as shown by its significance level of 78 per cent.

Table 8.1 Models fitted to data on perceptions of satisfaction with financial circumstances.

	Model*	G^2	Df	Significance (%)
(1)	Grand mean	26,235	299	0
(2)	S+F+I+A+E	5,331	287	0
(3)	S*F+S*I+S*A+S*E+F*I			
	F*A+F*E+I*A+I*E+A*E	280	234	2
(4)	S*F*I+S*F*A+S*I*A+F*I*A+S*F*E			
	F*A*E+S*I*E+F*I*E+S*A*E+I*A*E	111	123	78
(5)	Pairwise associations (model 3)			
(5.1)	plus S*F*I	263	218	2
(5.2)	plus S*F*A	259	226	6
(5.3)	plus S*I*A	267	226	3
(5.4)	plus F*I*A	246	202	2
(5.5)	plus S*F*E	244	230	26
(5.6)	plus F*A*E	254	226	9
(5.7)	plus S*I*E	275	230	2
(5.8)	plus F*I*E	255	213	2
(5.9)	plus S*A*E	276	232	3
(5.10)	plus I*A*E	269	226	3
(6)	S*F*E+S*I+F*I+S*A+F*A+I*A+I*E +A*E (model 5.5)			
(6.1)	minus S*I	256	234	16
(6.2)	minus F*I	787	246	0
(6.3)	minus S*A	847	232	0
(6.4)	minus F*A	1,096	238	0
(6.5)	minus I*A	254	238	22
(6.6)	minus I*E	627	234	0
(6.7)	minus A*E	411	232	0
(7)	S*F*E+F*I+S*A+F*A+I*E+A*E	266	242	14

* Model abbreviations:
S - satisfaction with current financial circumstances;
F - how family income compares with that of US families in general;
I - how family income when aged 16 compares with that of other US families at that time;
A - whether financial circumstances have improved or got worse in recent years;
E - level of education

However, in reaching this model, we have jumped from all associations to *all* interactions, neglecting to examine intermediate models including only one or a few interactions. One or more of these simpler models might also fit quite adequately. The next step in the forward selection procedure, therefore, is to see whether all the interaction terms in the model are really necessary. This is done by reverting to the pairwise-association model and finding the fit when just one of the interactions is added, for each interaction in turn.

Models 5.1–9 show the results: all the association-plus-one-interaction models have significance levels below 5 per cent, except models 5.5 and 5.6. Noting that model 5.5 has the highest significance level (26 per cent), we can conclude that all the interactions other than $S*F*E$ can be pulled out of model 4 (giving model 6) without destroying the fit. Model 6 is clearly much simpler than model 4, since it includes one rather than ten interactions. But it may still be possible to simplify it further by removing unnecessary associations. Because we are using only hierarchical models, the associations that are lower-order relatives of $S*F*E$ – that is, $S*F$, $S*E$ and $F*E$ – cannot be deleted. All the other associations are candidates for removal, and models 6.1–7 show the consequences of taking each of them out of model 6, one at a time. Removing both $S*I$ and $I*A$ individually leaves a model with a significance level well over 5 per cent; it seems that neither association is necessary in the model. Model 7 is the one that results from taking both these associations out of model 6, and is the final model obtained by forward selection.

To summarize what has been done, forward selection involves the following steps:

1 Try models including all "level L" effects successively, for L equal to 0 (that is, the grand mean model), equal to 1 (main effects), equal to 2 (associations) and so on, until a model is found that fits.

2 Return to the model before the one that fitted (that is, return to the level $L-1$ model). Examine all models, including any one level L effect; if none fit, examine those including any two level L effects; and so on, until a model is found that fits. (In the example above, a model was found that fitted after examining those including just one level L effect, but in other cases a model including two or more may be needed.)

3 Remove the $L-1$ level effects one at a time, until no more may be deleted without reducing the fit to an unacceptable degree.

The simplest model obtained by forward selection which fits the data is:

$$S{*}F{*}E + F{*}I + S{*}A + F{*}A + I{*}E + A{*}E$$

What can be inferred from this?

First, the model confirms that many of the indicators of perceptions of past and present financial circumstances are associated, as had initially been supposed. In particular, respondents' views on how their parents' family incomes compared with those of other families (*I*) are related to their views about their circumstances relative to other families now (*F*). Their educational level (*E*) is related to their views on how well off their parents were (*I*), perhaps because there is a tendency for the sons and daughters of rich parents to stay in education for longer than the poor. The respondents' assessments about whether their circumstances have got better or worse over the last few years (*A*) are related to their educational level (*E*), their current satisfaction with their finances (*S*) and their ideas about how their current circumstances compare with those of other families (*F*). The interaction term in the model indicates that the relationship between respondents' views on how satisfied they are with their income (*S*) and their ideas on how that income compares with that of other families (*F*) varies with their level of education (*E*). This may be because education tends to broaden people's horizons, so that the better educated are likely to compare their incomes not just with a small circle of people known to them but with a much wider reference group.

The forward selection method is a convenient technique for sifting through a large number of models to find one that is both relatively simple and that fits adequately. However, it is not without weaknesses. First, some models are never examined. For instance, a model including a pair of interactions and no associations, other than lower-order relatives, might turn out to be a better model than the one we selected; we do not know, because no models of this type were explored. Secondly, on some occasions the method yields not one, but two alternative models. For instance, in the example model 5.6 fits with a significance level of 9 per cent. We could have examined this model further, rather than choosing model 5.5. We might well have finished with a model not much more complex than model 7, which fitted the data just about as well, but which included a different interaction term. Forward selection has nothing to say about choosing between such alternative models. A decision has to be made using sociological knowledge about what sort of model seems most plausible. Thirdly, forward selection is a purely mechanical procedure which leaves no room for prior ideas and the analyst's goals to enter directly into the choice of a model. Nevertheless, despite these problems, it seems to locate a reasonable model in most cases.

Forward selection starts from the simplest model (the grand mean model) and progressively adds effects until a model that fits is found. Alternatively, we could have started with the most complex model (the saturated model, consisting of all possible effects) and gradually removed effects until a model is found that does not fit. At that point, we know that we have gone too far in simplifying and need to revert to the previous model. This alternative strategy is known as *backwards elimination*. There is not a lot to choose between the two strategies.

The coefficient of multiple determination

Although model 7 of Table 8.1 is the best model that forward selection can find, this does not mean that the model is necessarily particularly good at explaining the data. It would be useful to know just how good this "best" model is. There are, however, two quite different criteria that could be used in assessing the model. On the one hand, we could be interested in how well the data cell frequencies are fitted by the model. A measure will shortly be introduced which will allow this to be quantified and we shall find that model 7 fits the data cell frequencies very well. On the other hand, we could be interested in how completely the model explains people's level of satisfaction with their financial circumstances. This is really a question about whether all the relevant variables have been included in the model. Since only variables in the data table were tried in the model, we must also ask whether all relevant variables have been included in the data table.

The answer to this latter question must be no. As noted above, the level of respondents' actual income is an important determinant of level of satisfaction, and there are many other variables that could also be influential. In short, the data table tells some, but by no means all, of the story about levels of satisfaction. The part of the story it does tell is represented by the associations and interaction in the model we have fitted. However, *no* statistical measure based on the data table will help to decide how much of the complete explanation of satisfaction we have found. No measure will indicate whether we have included five relatively important or five relatively trivial variables in constructing the table. This means that the part of the story told by the variables omitted from the table may be, in fact, much more interesting and significant than the part we have found, but there is no way of telling whether this is so.

It is therefore vital to ensure that those variables that you think will be important are included in your data, and to justify the exclusion of other variables. This can be done only by recourse to the theoretical framework that underlies the research. With this in mind, we can now see how well the model we have selected fits the data. Remembering that a zero value of G-squared indicates that the data have been fitted perfectly, it is reasonable to interpret a non-zero value of G-squared as indicating the amount of variation in the data cell frequencies left unexplained by a model. A model that gives a fit with a G-squared value of zero explains all the variation in the data; a model with a large value of G-squared explains relatively little. This idea is used in defining the *coefficient of multiple determination,* which measures the proportion of the total variation in data cell frequencies explained by a model.

The coefficient is calculated by comparing the relative fit of the selected model with the fit of a "minimal" model including no effects, that is, with the grand mean model. Thus, the coefficient of multiple determination for model 7, which has a G-squared value of 266, is:

$$\begin{aligned}(G^2_{\text{grand mean}} - G^2_{\text{model 7}})/G^2_{\text{grand mean}} &= (26{,}235 - 266)/26{,}235 \\ &= 0.99\end{aligned}$$

meaning that the model has explained 99 per cent of the variation in the data table. One disadvantage of this coefficient is that for all but the very worst-fitting models its value is usually close to 100 per cent. For example, all the models 5.1–10 have a coefficient exceeding 99 per cent.

To conclude this chapter, let us use forward selection on the data of Table 8.2 from a well-known study by Stouffer et al. (1949) about the preference of Second World War recruits for the location of their training camp. Table 8.3 summarizes the results of forward selection. Note that models including only one interaction term seem to be insufficient to fit these data, since all such models failed to give a significance level above 5 per cent. Hence, it was necessary to examine models including two interactions. The only association that is not a lower-order relative of the best two-interaction model could not be deleted without destroying the fit. Make sure that you follow the logic of the choice of models to test shown in Table 8.3.

The final model (model 5.3) shows that preference for location of

Table 8.2 Preference of Second World War recruits for the location of their training camp.

Race (R)	Region of origin (O)	Location of present camp (L)	Number of recruits (P) preferring a camp in north	in south
	North	North	387	36
Black		South	876	250
	South	North	383	270
		South	381	1,712
	North	North	955	162
White		South	874	510
	South	North	104	176
		South	91	869

Source: adapted from Stouffer et al. (1949), with those undecided omitted.

Table 8.3 Forward selection on the data of Table 8.2.

	Model*	G^2	Df	Significance (%)
(1)	Grand mean	5,470	15	0
(2)	P + L + O + R	4,211	11	0
(3)	P*L + P*O + P*R + L*O + L*R + O*R	78	5	0
(4)	P*L*O + P*L*R + P*O*R + L*O*R	1	1	41
(4.1)	P*L*O + P*R + L*R + O*R	46	4	0
(4.2)	P*I*R + P*O + L*O + O*R	77	4	0
(4.3)	P*O*R + P*L + L*O + L*R	73	4	0
(4.4)	L*O*R + P*L + P*O + P*R	25	4	0
(5.1)	P*L*O + P*L*R + O*R	34	3	0
(5.2)	P*L*O + P*O*R + L*R	36	3	0
(5.3)	P*L*O + L*O*R + P*R	2	3	70
(5.4)	P*L*R + P*O*R + L*O	72	3	0
(5.5)	P*L*R + L*O*R + L*O	17	3	0
(5.6)	P*O*R + L*O*R + P*L	25	3	0
(6)	P*L*O + L*O*R	153	4	0

* Variable abbreviations:
P – preferred location of camp;
L – present location of camp;
O – region of origin;
R – race

training camp (variable *P*) depends on race (*R*), region of origin (*O*),location of present camp (*L*) and the interaction between region of origin and location of present camp. The latter assertion amounts to saying that the preferred location depends on where the respondent is at present, but the strength of this relationship varies according to his original place of residence. Moreover, there is an interaction between present location, race and region of origin; that is, the location of the respondent varies according to his region of origin, but more or less strongly depending on the respondent's race. The influence of race on the association between region of origin and present location may be a sign of racial discrimination in the allocation of recruits to training camps.

In this chapter we have shown how a model can be found to fit a complex table. The models that were chosen involved both associations and interactions. The next question that is likely to be asked is whether all the relationships in a model are equally important. It may be that one or more of them are very weak, having little substantive significance. What is needed, therefore, is a way of measuring the strengths of effects. In Chapter 4 we discussed measuring the strength of an association in a two-dimensional table; in Chapter 9 we review measures that can be used for more complex tables.

Summary

Although the analysis of higher-dimension tables is similar to that of tables involving only three variables, the additional dimensions do introduce two complications. In the first place, second- and third-order interactions may be found, although in practice these occur infrequently. Secondly, a large number of models may be fitted to such tables, making a strategy for finding the best necessary. There are two standard strategies, forward selection and backwards elimination. However, sometimes even the "best" model may actually fit the data rather poorly. Whether this is the case can be assessed with the coefficient of multiple determination.

Computer analysis

Model selection using forward selection and backwards elimination are in principle mechanical procedures, just involving following the rules set out above. They can therefore be automated. SPSS offers backwards elimination, while GLIM provides a shorthand way of specifying models which makes either model selection procedure relatively easy to carry out. With both programs, however, you need to be careful about accepting uncritically the chosen model, because there may be other more appropriate models which have not been examined.

SPSS

The SPSS command, HILOGLINEAR will select a model using backwards elimination automatically, without your having to specify each model to test. Program 8.1 reads the data of Table 3.2 about pre- and extra-marital sex and finds the best model using backwards elimination.

Program 8.1

```
DATA LIST FREE / MarStat, EMS, PMS, Gender, Count.
BEGIN DATA.
1 1 1 1  17
1 2 1 1  54
1 1 2 1  36
1 2 2 1 214
1 1 1 2  28
1 2 1 2  60
1 1 2 2  17
1 2 2 2  68
2 1 1 1   4
2 2 1 1  25
2 1 2 1   4
2 2 2 1 322
2 1 1 2  11
2 2 1 2  42
2 1 2 2   4
2 2 2 2 130
END DATA.
VARIABLE LABELS
 MarStat 'Marital Status'
 EMS 'Extra-marital Sex'
 PMS 'Pre-marital Sex'.
VALUE LABELS
 MarStat 1 'Divorced' 2 'Married'/
 EMS, PMS 1 'Yes' 2 'No'/
 Gender 1 'Women' 2 'Men'.
WEIGHT BY Count.
HILOGLINEAR Gender(1,2), PMS(1,2), EMS(1,2), MarStat(1,2)
```

```
/METHOD=BACKWARD
/MAXORDER=4
/PRINT=NONE.
```

The HILOGLINEAR command begins by fitting the second-order interaction between all four variables (which of course fits perfectly), and then proceeds to test all interactions, all associations, all main effects and the grand mean models. It finds that the second-order interaction is not needed and then tries deleting each interaction from an all-interaction model, one at a time. Fitting is continued in this fashion until the best model is located:

PMS*EMS*MarStat + Gender*PMS + Gender*EMS + Gender*Marstat

This has a *G*-squared of 0.76 with 4 degrees of freedom, giving a significance level of 94 per cent, an excellent fit.

GLIM

Unlike SPSS, GLIM does not have any automatic means of carrying out forward selection or backwards elimination, but it is possible to fit a sequence of models, one at a time. To assist with this, GLIM allows models to be specified by indicating only the effects to be added to or deleted from the model previously fitted. This is often much quicker than re-entering the complete model. Some care has to be taken to add and delete the right terms. The GLIM convention is to use an asterisk (*) in the term for an effect *including* the lower-order marginals and a full stop for the effect alone. The distinction is important when it comes to deleting effects from a model. For instance, the GLIM directive

$fit *A**B**C*

fits the model $A + B + C + A.B + A.C + B.C + A.B.C$. To change to fitting a model without the interaction term, use the directive

$fit −*A.B.C*

leaving the model $A + B + C + A.B + A.C + B.C$. If instead you were to use the directive

$fit −*A**B**C*

there would be nothing left in the model, since not only the interaction, but also all its lower-order relatives would have been deleted.

In Program 8.2, the variables' names have been abbreviated to their initial letters to make it quicker to enter complex model specifications. In this program, first the grand mean model (specified as $fit 1), and then the main effects, the associations and the interactions models are fitted. The interactions model fits very well (*G*-squared = 0.15 with 1 degree of freedom), so we return to the associations model and add one interaction term. Because this does not fit well, that interaction is deleted and another tried. Neither this nor the next interaction yields a well fitting model, but the fourth does (*G*-squared = 0.76, 4 d.f.). We now have a model with all the associations plus one interaction. It may be that some of the associations are superfluous, so we try deleting them in turn. Deleting *G.P* makes the *G*-squared rise sharply, to 70.1, so it is replaced, and *G.E* is deleted. This has little effect on the *G*-squared, nor does the removal of *G.M*. Other associations cannot be deleted because they are lower-order relatives of the interaction term, *P.E.M*, which has already been shown to be necessary in the model.

Program 8.2

```
$units 16
$factor M 2 E 2 P 2 G 2
$calc M = %GL(2,8)  : E = %GL(2,1)  : P = %GL(2,2)  : G = %GL(2,4)
$data Count
$read
  17  54  36 214
  28  60  17  68
   4  25   4 322
  11  42   4 130
$yvar Count
$error p
$fit 1
$fit G + P + E + M
$fit G*P + G*E + G*M + P*E + P*M + E*M
$fit G*P*E + G*P*M +G*E*M + P*E*M
$fit G*P + G*E + G*M + P*E + P*M + E*M + G.P.E
$fit - G.P.E + G.P.M
$fit - G.P.M + G.E.M
$fit - G.E.M + P.E.M
$fit - G.P
$fit + G.P - G.E
$fit - G.M
```

The result of this process is that the best model is found to be $P{*}E{*}M + G{*}P$, which has a *G*-squared of 8.15 with 6 degrees of freedom and a significance level of 23 per cent.

The forward selection strategy has yielded the model

$$P{*}E{*}M + G{*}P$$

but backwards elimination using SPSS, on the same data, resulted in the model

$$P{*}E{*}M + G{*}P + G{*}E + G{*}M$$

The former fits well (at the 23 per cent level) and is simpler than the latter, which also includes the $G{*}E$ and $G{*}M$ associations and fits at the 94 per cent level. The reason that the backwards elimination included these extra associations is that each is individually statistically significant, although in practical terms neither represents a strong association, and thus removing them from the model still leaves it fitting well. In the next chapter we shall see how the strength of these individual effects can be measured.

Further reading

Bishop, et al. (1975) deal with a variety of methods of model choice, including forward selection and backwards elimination, and discuss their relative merits. See also Benedetti & Brown (1978). The coefficient of multiple determination was introduced by Goodman (1972).

CHAPTER NINE

Effects and odds ratios

In the previous chapter, we saw how even complex multidimensional tables can be modelled in terms of a number of associations, interactions and other effects. In this chapter, attention will turn to unpacking these effects to examine the specific influence of one category of a variable on others. Before doing this, however, we must set the material covered so far on a firmer foundation, by examining more closely the link between the counts in the data table and the effects in a model. The first part of the chapter reviews why it is important to consider not just the statistical significance of effects, but also their magnitude. Next, we develop an equation for a loglinear model to fit a simple 2 × 2 table which will incidentally explain why loglinear analysis is so called. The loglinear equation will show that the relationships between variables can be measured in terms of the "odds ratio" described in Chapter 4.

In Chapter 8, the simplest model that fitted data on satisfaction with standard of living had six terms, each of which was shown to be statistically significant. However, the fact that all six were needed in the model does not mean that they all represented substantively important relationships. It may be that particular terms correspond to relationships that, although present in the data, are nevertheless so small in magnitude that they can be neglected as trivial in any sociological analysis. In order to see whether this is so, it is necessary to assess the magnitude of a relationship.

The distinction between a statistically significant and a substantively significant effect is an important one. Suppose that in some population 40 per cent of males and 41 per cent of females believe in the existence of God. Clearly, the relationship between this religious belief and gender is of no sociological importance whatsoever: no researcher would worry about explaining such a very small difference between men and women. Table 9.1 shows the results one might get from a survey of 100 males and 100 females drawn from this population. Fitting a model of no association to the table yields a G-squared of 0.02 which, with 1 degree of freedom, gives a significance level of 88 per cent - no association between the variables has been detected. Now consider Table 9.2, showing the results from a much more ambitious survey of 200,000 individuals. Although the table is based on exactly the same population, and again precisely 40 and 41 per cent of males and females state that they believe in God, fitting a model of no association gives a G-squared equal to 20.7 and a significance level of 0.01 per cent, showing that there *is* association in this table.

Table 9.1 Belief in God by gender: invented data from 200 respondents.

	Male	Female
Believes in existence of God	40	41
Does not believe in God	60	59

Table 9.2 Belief in God by gender: invented data from 200,000 respondents.

	Male	Female
Believes in existence of God	40,000	41,000
Does not believe in God	60,000	59,000

The moral to be drawn from this is that the power of loglinear analysis to detect a relationship depends not only on the strength of the effect, but also on the sample size. Given a large enough sample, even entirely trivial effects may be detected. Conversely, the fact that an effect must be included in a model for it to fit does not mean that the effect is of any practical importance. More than knowledge of significance levels is needed to measure the strengths of relationships.

It is also the case that in modest-sized samples sociologically important effects may not be statistically significant and so may be missed. You should be on your guard that with very large samples (of several thousand respondents) you do not draw sociological conclusions about effects that are of little substantive importance, and that with small samples (of a few hundreds of respondents) you do not assume that because an effect is not needed for a fitting model, the effect is not present in the population.

As the illustration above shows, enlarging the sample size will affect the value of measures of significance, because they are dependent on the absolute magnitude of differences in frequencies within a data table. Measures of association are not sensitive to the actual magnitude of the data frequencies, but depend on the proportions of the sample that fall into each cell. Changing the sample size should, therefore, have no effect on measures of association. Similarly, measures that indicate the effect of particular categories of variables on cell counts are insensitive to sample size. It is to defining such measures in terms of the basic loglinear equation that we now turn.

The loglinear equation

As was shown in earlier chapters, a two-dimensional table may include one or more of four effects: the grand mean, the main effect of one variable (say, variable *A*), the main effect of the other variable (variable *B*), and the association between the two. An equation can be constructed which sets the individual cell counts in the model table equal to the sum of four *parameters* that are based on measures of these four effects. We will call these parameters "μ-terms". (μ is the Greek letter m and is pronounced "mu".)

The derivation of the loglinear equation will be illustrated with data from part of a study of the US International Typographical Union. Table 9.3 cross-tabulates the number of printing workers on day or night shift with whether they ever visited the homes of other printers. Lipset et al. (1956), from whom this table is taken, suggest that, although night-shift printers generally associated more with other printers, the disruption of normal family schedules resulting from night work led to fewer of them making home visits than was the case amongst day-shift printers.

Table 9.3 Effect of night working by printers on visiting other printers at home.

Variable A	Variable B	
Do you ever visit other printers at home?	Night workers	Day workers
Yes	141	178
No	58	56

Source: Lipset et al. (1956: Table 13).

The grand mean of the table measures the average count in each cell, but is based on a geometric rather than an arithmetic mean. It is calculated by first multiplying together the frequencies in all the cells, then taking a root of the result. For instance, for a 2 × 2 table, we take the fourth root of the product of the four cell frequencies:

$$\sqrt[4]{x_{11}x_{12}x_{21}x_{22}} = (x_{11}x_{12}x_{21}x_{22})^{1/4}$$

Note that taking the fourth root is equivalent to taking the quarter power. This expression is rather inconvenient, since it involves products and roots. It can be converted into a more convenient form by taking its logarithm. The advantage of logarithms is that products of numbers are transformed into sums of logarithms, division becomes subtraction, and raising one number to the power of another is converted to multiplication of the two numbers. By convention, natural logarithms, which are based on the constant e (2.7183), are always used. The usual symbol for a natural logarithm is either "ln" or "$\log_e$" to distinguish it from the logarithms found in log tables (which are logs to the base 10), but since all the logarithms in this book are natural logarithms, we shall just use the symbol "log".

Taking logarithms of the expression for the grand mean gives the grand mean μ-term:

$$\mu = \frac{1}{4}(\log x_{11} + \log x_{12} + \log x_{21} + \log x_{22})$$

which is just the average of the logarithms of the cell frequencies. This equals 4.554 for the data of Table 9.3.

Let us leave the μ-terms for the main effects for a moment, and deal next with the μ-term for association. The magnitude of an association can be measured with the odds ratio which was mentioned in Chapter 4, equal to:

$$(x_{11}x_{22})/(x_{12}x_{21})$$

This divides the product of the top left and bottom right cell counts by the product of the top right and bottom left counts. Taking the fourth root and logs gives a μ-term representing the association between the two variables, A and B, for cell (1,1):

$$\begin{aligned}\mu_{11}^{AB} &= \log\left(\sqrt[4]{(x_{11}x_{22})/(x_{12}x_{21})}\right)\\ &= \frac{1}{4}(\log x_{11} - \log x_{12} - \log x_{21} + \log x_{22})\end{aligned}$$

which, for the data of Table 9.1, equals −0.067.

In fact, this is just one of four related association μ-terms. The odds ratio might have been computed by starting at the bottom right (x_{22}) instead of the top left (x_{11}) corner, giving the same result but a formula with the subscripts interchanged. Starting from either of the other two cells gives the reciprocal of the odds ratio. Corresponding to these four ways of finding the odds ratio are four association μ-terms, known generically as μ_{ij}^{AB}, where for a 2 × 2 table both i and j can take the values 1 or 2.

Now we can deal with the main-effect terms. The main effect of a two-level variable is the geometric mean of the frequencies in one category relative to the frequencies in the other category. For variable A, this is the product of the top left and the top right cell counts, divided by the product of the bottom left and bottom right counts:

$$(x_{11}x_{12})/(x_{21}x_{22})$$

Note the different arrangement of the counts in this expression and in the odds ratio above.

The main-effect μ-term is the log of its fourth root, and for category 1 of the variable A is:

$$\mu_1^A = \frac{1}{4}(\log x_{11} + \log x_{12} - \log x_{21} - \log x_{22})$$

The main effect for category 2 of this variable is the same, but "upside down" – the mean of the frequencies in category 2 relative to those in category 1. The equation for μ_2^A is therefore the same as for μ_1^A, but with the signs between the logs interchanged. Consequently,

$$\mu_2^A = -\mu_1^A$$

For the "visiting" variable of Table 9.1, μ_1^A is equal to 0.511. This is the value for the main effect of those visiting printers at home; for those not visiting, μ_2^A, the value is −0.511. The main-effect μ-term for variable B (shift working) can be expressed similarly and evaluates to −0.495 for night shift working, and 0.495 for the day shift.

Another way of looking at the main effect is that it is the difference between the average of the log counts for a particular category and the grand mean. A positive main-effect μ-term results when the geometric mean of cases in a category is greater than the mean for the whole table. For instance, the mean number of printers saying that they do visit other printers ($\sqrt[2]{141 \times 178}$) is greater than the mean cell count ($\sqrt[4]{141 \times 178 \times 58 \times 56}$) and so μ_1^A is positive.

Finally, all these μ-terms can be put together to form a loglinear equation for the saturated model (the one that includes all effects):

$$\log m_{ij} = \mu + \mu_i^A + \mu_j^B + \mu_{ij}^{AB}$$

The model frequency m_{ij} enters the equation in the log form because logs have been taken to express the μ-terms on the right-hand side. The equation means that, if the values of the μ-terms for a particular table are calculated using the formulae given above, they can be summed to yield the log of the frequencies that should go in the model table. The equation also, incidentally, reveals why loglinear analysis is so called: the equation yields a *log* frequency and is *linear* (that is, it contains sums of simple terms).

This equation is rarely used for calculating model tables. Nevertheless, let us check the μ-term values derived above by obtaining the model cell count for the top left-hand corner cell:

$$\begin{aligned}\log m_{11} &= \mu + \mu_1^A + \mu_1^B + \mu_{11}^{AB} \\ &= 4.554 + 0.511 - 0.050 - 0.067 \\ &= 4.948 \\ m_{11} &= 140.9\end{aligned}$$

The equation has regenerated (almost) the original data frequency. The result is not exactly right only because of rounding errors in the calculations. Application of the equation will yield model frequencies almost exactly equal to the data cell frequencies because the saturated model is being used – that is, the model with all possible terms included. The way that μ-terms can be combined to yield the other model cell counts can be seen from Table 9.4. (Remember that $\mu_2^A = -\mu_1^A$ and $\mu_2^B = -\mu_1^B$.)

Table 9.4 Relationship between the logarithms of model table counts and μ-terms.

Variable *A*	Variable *B*	
	1	2
1	$\mu + \mu_1^A + \mu_1^B + \mu_{11}^{AB}$	$\mu + \mu_1^A - \mu_1^B - \mu_{11}^{AB}$
2	$\mu - \mu_1^A + \mu_1^B - \mu_{11}^{AB}$	$\mu - \mu_1^A - \mu_1^B + \mu_{11}^{AB}$

To generate a model table showing no association, the full equation is adapted by omitting the association μ-term:

$$\log m_{ij} = \mu + \mu_i^A + \mu_j^B$$

Using this equation would lead to the same model table as would be obtained by fitting the model, $A*B$.

There is a formal one-to-one correspondence between the terms in a loglinear model equation and the set of effects in a model. It is this correspondence that allowed us to describe a model in terms of its marginals without needing to know either the underlying equation or the values of the μ-terms. With higher-dimension tables, further μ-terms corresponding to interactions and higher-order effects become involved.

Equations for any particular model can be written down by including one term for each effect in the model. However, not all such equations can be solved using analytic (that is, algebraic) methods. For instance, the equation for a pairwise-association model cannot be solved analytically. For such equations the only method of solution is an iterative one such as the iterative proportional scaling procedure described in Chapter 5.

The program that carries out the iterative procedure will also calculate the values of all the μ-terms. Once these μ-terms have been obtained they can be used to measure the strengths of effects, either directly or by calculating odds ratios from them. Odds ratios are preferred because it is usually easier to provide an interpretation of an odds ratio than a μ-term. The conversion from a μ-term for an association such as μ_{ij}^{AB} to an odds ratio is straightforward: as noted above, μ_{ij}^{AB} is a quarter of the logarithm of the odds ratio and so the odds ratio is the exponential of four times μ_{ij}^{AB}.

Odds ratios are therefore important in interpreting the strengths of effects in loglinear models. In the next section, a study that used odds ratios to examine patterns of relationship will be described.

Interpreting models with odds ratios

In 1989 the UK General Household Survey asked a national representative sample of respondents whether they had co-habited with their partners before marriage, for each marriage in which they had been involved. The data in the Survey also indicate whether the respondents' first marriages had ended in divorce by 1989. These data allow the investigation of whether living with a partner before marriage is associated with more successful marriages (that is, ones less likely to end in divorce). In an interesting study by Haskey (1992), it is shown that the reverse appears to be the case: those who cohabit with their future spouse before marriage are *more* likely to divorce than those who do not.

Haskey breaks down the sample as a whole into groups by sex, by age (in ten-year bands) and by age at first marriage, on the assumption that these factors will affect the likelihood of divorce after cohabitation. He then examines odds ratios for each group separately. (An alternative would have been to fit a model that included the variables marital status, cohabitation, age, sex and age at first marriage, but we will follow Haskey's treatment of the data here.) Table 9.5 shows the observed counts for the group of couples in which the husband married during the 1950s aged between 20 and 24. A model of no association fails to fit (G-squared = 11.2 with 1 degree of freedom), indicating that there is a relationship between premarital cohabitation and the marriage ending in divorce. The strength of the relationship can be measured by fitting a saturated model and estimating the μ_{11}^{AB} parameter, which is equal to 0.20.

Table 9.5 First marriages, premarital cohabitation and subsequent divorce among men married in the 1950s when aged 20–24.

	Ended in divorce	Did not end in divorce
Pre-marital cohabitation	36	80
No pre-marital cohabitation	94	462

Source: Haskey (1992: 14).

To explain what this result means, it is best to express it in terms of the odds ratio. Because μ_{11}^{AB} is a quarter of the natural logarithm of the odds ratio, the odds ratio can be calculated to be 2.21. Slightly more than twice as many marriages ended in divorce for couples who cohabited before marriage compared with those who did not. Another way of expressing the same fact is that there is a 2.2 greater risk of divorce among those who cohabited before marriage than among those who did not (considering only couples who married in the 1950s where the husband was aged 20–24 at the date of the marriage).

In making these kinds of statements, we must be careful not to imply that, because there is an association between the variables, the link is a causal one. Just because cohabitees are more likely to divorce does not mean that cohabitation causes divorce. There may be other factors common to both cohabitation and divorce. For example, Haskey (1992: 17) suggests that those who cohabit premaritally might have a lesser belief in marriage as a life-long commitment. It is also possible that those who were unsure about their partners may have decided to cohabit for a period before marriage to test the relationship; the greater probability of the subsequent breakdown could be associated more with the initial doubts than the cohabitation itself. These alternative explanations for the relationship suggest why the finding of an association between variables does not automatically show that they are causally related.

The odds ratio applies to the specific sample of 672 couples in Table 9.5. In order to know whether the relationship between cohabiting and divorce is likely to be characteristic of the population as a whole, it would be useful to know the *standard error* of the value of μ_{11}^{AB}. If a number of different samples are taken from the UK population, it would be expected that the value of μ_{11}^{AB} calculated from each sample would differ slightly between samples because of random variations in who is chosen for inclusion in the sample. The standard error indicates the amount of sampling variablity that could be expected. The lower the standard error, the more certain it is that the estimate from a sample is close to the true population value.

If the model fits the data, the sample is random and the sample size is sufficiently large, the ratio of the value of the estimate and its standard error is normally distributed with mean 0 and variance 1. This fact can be used to construct a *confidence interval* around the estimate. The confidence interval is the band of values within which it is 95 per cent probable that the true population value lies. The lower bound of the confidence interval is 1.96 multiplied by the standard error less than the estimate, and the upper bound is 1.96 multiplied by the standard error more than the estimate.

Computer programs that calculate loglinear parameter values also provide the standard errors (and, usually, the confidence intervals). For example, the standard error for the μ_{11}^{AB} parameter is 0.057. Recalling that the value of μ_{11}^{AB} is 0.20, the upper and lower bounds of the confidence interval around μ_{11}^{AB} are

$$0.20 - 1.96 \times 0.057 = 0.09$$
$$0.20 + 1.96 \times 0.057 = 0.31$$

This means that there is a 95 per cent probability that the true value of this parameter in the population as a whole is between 0.09 and 0.31. In particular, there is little chance that the population value is zero (because zero does not lie within this confidence interval), thus confirming that there really is a relationship between cohabiting and divorce and that it is rather unlikely that the relationship is merely a statistical freak arising from the character of the specific sample that contributed to Table 9.5.

Parameters for more complex tables

In the previous sections, we have only considered "2 × 2" tables to simplify the discussion. For these tables, there can only be grand mean, main-effects and association terms in the loglinear equation. If there are more than two variables, further μ-terms, one for each of the effects in the model, need to be included. If some of the variables have more than two levels, more parameters will be needed for each effect.

Table 9.6 The domestic division of labour: doing the dishes.

Who does the evening dishes in your household?*	Man works, woman works full-time	Man works, woman works part-time	Man works, woman does not work	Other, not answered
Mainly man	45	36	38	99
Mainly woman	68	76	150	87
Shared equally	84	68	95	108

* I would like to ask about how you and your [husband/wife/partner] generally share some family jobs. Who does the evening dishes: mainly the man, mainly the woman, or is the task shared equally?
Source: derived from Jowell et al. (1988: 198).

For example, when it comes to variables such as the division of domestic labour in Table 9.6, the main-effect μ-term consists of three values, one for each category (see Table 9.7). These indicate the degree to which each category has an average log count greater than the grand mean. Because if some categories have values greater than the average, others must have values less than the average, the values that make up μ-terms always sum to zero.

The association μ-term is also more complicated than in case of a "2× 2" table. There is one value for each cell in the table (see Table 9.7c). Excluding the "Other" column, the highest value, 0.40, is for the cell, "Mainly woman does dishes/Man works, woman does not work". This means that there are more respondents reporting this specific division of domestic labour than would be expected if who did the dishes were independent of work status. In other words, in households where women do not work, they do the dishes more often, and men less often, than equality would seem to warrant. Perhaps more unexpectedly, in households where both partners are in full-time work, there is some tendency for the evening dishes to be shared out more equally and for the woman to do the washing up less often than might be predicted from the assumption that who does the dishes is independent of gender.

Table 9.7 Parameters from fitting a saturated model to Table 9.6.

(a) The values of μ_i^A

Mainly man	−0.39
Mainly woman	0.21
Shared equally	0.18

(b) The values of μ_j^B

Man works, woman works full-time	−0.14
Man works, woman works part-time	−0.25
Man works, woman does not work	−0.11
Other, not answered	−0.28

(c) The values of μ_{ij}^{AB}

Who does the evening dishes in your household?	Man works, woman works full-time	Man works, woman works part-time	Man works, woman does not work	Other, not answered
Mainly man	*0.04*	*−0.07*	−0.37	*0.40*
Mainly woman	−0.14	*0.07*	0.40	*−0.32*
Shared equally	*0.10*	*0.00*	*−0.02*	*0.08*

Note: Values not significantly different from zero at the 5 per cent level shown in *italic*.

We shall return to the use of μ-terms and odds ratios to measure effects in the next chapter, which is concerned with problems where the aim is to predict or explain the values of one variable from the values of others using a form of loglinear model known as logistic regression.

Summary

Loglinear models are so called because they are based on an equation that sets the natural logarithm of the model table frequencies equal to sums of terms, one for each marginal that is included in the model. The values of these terms and their standard errors can be calculated by the computer programs used to fit models and may be used to assess the magnitude of the effects in the table.

Computer analysis

Both SPSS and GLIM will calculate parameters for loglinear models, but the coefficients they provide differ. Unless instructed otherwise, SPSS calculates the values of the μ-terms themselves. As noted above, main-effect μ-terms can be interpreted as the deviations around the grand mean of a table. It is also possible to calculate the main-effect parameter of a variable in terms of deviations from one of the variable's categories, selected as a "reference" category. This is the choice adopted by GLIM. In both cases, the same information is available and the coefficients can be transformed from one form to the other with a little arithmetic.

SPSS

The SPSS LOGLINEAR procedure can be used to calculate the parameters of the loglinear equation for a saturated model fitted to the data about printers of Table 9.3 (see Program 9.1). The /PRINT ESTIM subcommand requests the parameters to be displayed, as in Table 9.8.

Program 9.1

```
DATA LIST FREE /Visit, Night, Count.
BEGIN DATA.
1 1 141
1 2 178
2 1  58
2 2  56
END DATA.
VARIABLE LABELS Visit 'Visit other printers at home?'/
  Night 'Is a night worker'.
VALUE LABELS Visit 1 'Yes' 2 'No'/
  Night 1 'Night worker' 2 'Day worker'.
WEIGHT BY Count.
LOGLINEAR Visit(1,2), Night(1,2)
 /PRINT ESTIM
 /CRITERIA=DELTA(0)
 /DESIGN = Visit, Night, Visit BY Night.
```

Loglinear analysis programs sometimes have trouble fitting models to data that include "empty" cells, that is cells with zero counts. To avoid such problems, SPSS normally adds 0.5 to every cell count before fitting saturated models. For large samples, this makes little difference to the values of the parameters. In this case, however, to make the results from the SPSS and GLIM programs (see Program 9.2) more easily comparable, the addition has been suppressed using the /CRITERIA= DELTA(0) subcommand.

Table 9.8 Part of the output from SPSS for the data of Table 9.3.

```
Estimates for Parameters
VISIT
Parameter  Coeff.         Std Err.  Z-Value  Lower     Upper
                                             95 CI     95 CI
1          .5111871848    .05466    9.35161  .40405    .61833

NIGHT
Parameter  Coeff.         Std Err.  Z-Value  Lower     Upper
                                             95 CI     95 CI
2          -.0494830850   .05466    -.90524  -.15662   .05766

VISIT BY NIGHT
Parameter  Coeff.         Std. Err. Z-Value  Lower     Upper
                                             95 CI     95 CI
3          -.0670287449   .05466    -1.22622 -.17417   .04011
```

The SPSS parameter coefficients ("Coeff.") are equal to the values for the μ-terms calculated above. The output from SPSS omits the last coefficient for each parameter, but it can be calculated from the fact that all the coefficients sum to zero. For example, because the value for the first category of the main-effect parameter of the "Visit" variable is 0.51 and there are just two categories, the value for the second category must be -0.51. SPSS does not calculate the values of the grand mean parameter.

The Z-value shown in Table 9.8 is obtained by dividing the coefficient by the standard error. If the absolute value of Z exceeds 1.96, the coefficient can be considered to be significantly different from zero at the 5 per cent level. Notice that parameter 2 (for the main effect of the variable Night) has a Z-value less than 1.96 and the confidence interval includes zero. This indicates that this main effect is not statistically significant.

GLIM

Program 9.2 fits a saturated model to the data of Table 9.3. The

"$display e" directive requests that GLIM calculate the estimates of the parameters.

Program 9.2

```
$units 4
$data Visit Night Count
$read
1 1 141
1 2 178
2 1  58
2 2  56
$factor Visit 2 Night 2
$yvar Count
$error p
$fit Visit*Night
$display e$
```

Table 9.9 shows part of the output from this program.

Table 9.9 Part of the output from GLIM for the data of Table 9.1.

	estimate	s.e.	parameter
1	4.949	0.08422	1
2	-0.8883	0.1560	VISI(2)
3	0.2330	0.1127	NIGH(2)
4	-0.2681	0.2187	VISI(2).NIGH(2)

The estimate of the grand mean parameter is printed on the line labelled "1" under the heading "parameter" and is equal to the logarithm of the model table count for the cell corresponding to the first category of all the variables.

The parameter values for the first category of the other terms in the model are always set to zero and GLIM does not print them. So, for example, the estimate for the first category of the Visit variable (VISI) is 0 and for the second category is -0.8883.

These estimates are not the values of the μ-terms described above and calculated by SPSS. Nevertheless, they are directly related to μ-terms and can be used to calculate the μ-term values. The difference between the GLIM parameters and μ-terms arises from GLIM's slightly different method for expressing the basic loglinear equation. Instead of the relationship between model table counts and μ-terms shown in Table 9.4, GLIM uses the arrangement shown in Table 9.10, where the α_i are the four parameters numbered 1 to 4 in Table 9.9.

Table 9.10 The relationship between logarithms of model table frequencies and GLIM parameters.

Variable A	Variable B	
	1	2
1	α_1	$\alpha_1 + \alpha_3$
2	$\alpha_1 + \alpha_2$	$\alpha_1 + \alpha_2 + \alpha_3 + \alpha_4$

Recalling that the grand mean term, μ, is the average of the log counts,

$$\mu = \frac{1}{4}[\alpha_1 + (\alpha_1 + \alpha_2) + (\alpha_1 + \alpha_3) + (\alpha_1 + \alpha_2 + \alpha_3 + \alpha_4)]$$

$$= \alpha_1 + \frac{\alpha_2 + \alpha_3}{2} + \frac{\alpha_4}{4}$$

and, similarly, using the expressions for deriving the main effects and association terms from the model cell frequencies,

$$\mu_1^A = -\frac{\alpha_2}{2} - \frac{\alpha_4}{4}$$

$$\mu_{11}^{AB} = \frac{\alpha_4}{4}$$

GLIM also reports the standard error (s.e.) of each α_i, and these can be used to see whether they are significantly different from zero.

The advantage of GLIM's presentation of parameters is that effects relative to the first category of each variable can be obtained directly. For example, as shown in Table 9.10, for those who are night workers the log odds of visiting to not visiting is −0.8883, equivalent to an odds of 0.41. That means that there are 0.41 fewer visiting night workers than night workers who do not visit. The log odds ratio of visiting to not visiting, for night workers compared with day workers, is given as −0.2681, equivalent to an odds ratio of 0.76. In other words, the consequences of the disruption of normal family schedules resulting from night work led to fewer night-shift printers making home visits than day-shift workers, by a factor of 0.76.

Further reading

Bishop et al. (1975), Fienberg (1980), Haberman (1978), Reynolds (1977) and Agresti (1990) all develop the equations introduced in this chapter. Reynolds is probably the easiest to begin with. The idea of using standard errors and confidence intervals to make inferences about population parameters is introduced in the context of regression models by most introductory statistics textbooks, for example Blalock (1979). Wrigley (1985: 132–6) compares the coding systems used to generate the SPSS form of parameter coefficients with the GLIM form.

CHAPTER TEN
Logistic regression

In all the analyses we have looked at so far, the aim has been to model the pattern of cell counts within a table. However, there are many examples of data where one variable is of special significance. For these data, the object is not to explain the patterns of cells counts, but to explain or predict the values of this variable in terms of the remaining variables. For example, data cross-tabulating whether the respondent smokes cigarettes by age, sex and education might be used to try to explain smoking in terms of these other variables. Here, 'smoking' is the *dependent* or *response* variable, whose values are to be explained by the other variables, the *independent, predictor* or *explanatory* variables.

This chapter will describe a type of loglinear model, called a logistic model, which is intended for the analysis of data where there is one dependent variable which is to be explained by other variables. In the following chapter, we shall see how these models can be used to test explanations that explicitly suggest that one or more variables have a causal influence on others.

Table 10.1 shows data from a study of US politics which examined some of the influences on voting for Senators in statewide elections (Cowart 1973). Cowart chose three variables to cross-tabulate with voting either Democratic or Republican. The first was the voters' basic party identification - essentially, whether they thought of themselves as Democrats, Republicans or Independents. The second variable indicated whether any of the candidates was an incumbent, that is, the current Senator fighting to retain his or her position. This variable had three categories: one of the candidates was a Democratic incumbent, one of the candidates was a Republican incumbent, or no candidate was an incumbent. The third variable concerned voters' attitudes towards the parties' policies on domestic issues and towards the parties as managers of government. The data were obtained from a series of four surveys of US national samples over the period 1956–68.

Table 10.1 Voting for US Senators, by incumbent's party, attitude to party performance and party identification.

Party identification	Attitude	Incumbency	Vote D	Vote R
D	D	D	288	25
D	D	No	139	18
D	D	R	193	37
D	N	D	185	18
D	N	No	83	15
D	N	R	99	31
D	R	D	49	20
D	R	No	29	14
D	R	R	38	29
I	D	D	48	13
I	D	No	14	6
I	D	R	27	21
I	N	D	81	51
I	N	No	20	29
I	N	R	50	70
I	R	D	30	51
I	R	No	18	39
I	R	R	21	63
R	D	D	7	19
R	D	No	2	8
R	D	R	4	28
R	N	D	20	63
R	N	No	13	49
R	N	R	19	88
R	R	D	23	155
R	R	No	13	129
R	R	R	14	227

Key: D - Democrat, I - Independent, R - Republican, N - neutral, No - no incumbent.
Source: Cowart (1973: Table 2).

Although the kinds of model developed in previous chapters could be fitted to the data in this table, those models focus on explaining the cell counts, rather than the values of any particular variable. The natural questions to ask of these data is whether party identification, respondents' political attitude or incumbency make a difference to whether respondents vote Democratic or Republican and, if they do, what effect these variables have on voting. Vote therefore needs to be treated as a dependent variable and the others as explanatory variables.

LOGISTIC REGRESSION

We begin with models for data where the dependent variable is dichotomous, that is, has just two values. This is the case for the variable Vote, where the only possible choices tabulated are Democrat and Republican. Suppose that we choose to represent a vote for the Democrats with a code of 0 and a vote for the Republicans with a code of 1. Each respondent would thus be coded as 0 or 1 on the Vote variable and could be coded similarly 0, 1 or 2 on each of the other variables. If the problem is set up in this way, it may seem possible to examine the relationship between Vote and the explanatory variables using multiple regression, a technique that is also designed to explain one variable in terms of other variables.

There are two difficulties with using multiple regression in this case, however. First, the dependent variable can take only two values (0 and 1, representing Democrat or Republican). Multiple regression assumes that the dependent variable is continuous and can in principle take any value. There is a danger that a model based on multiple regression would predict that a certain respondent's score on the Vote variable is, for example, 0.5, and there is no way to interpret such a prediction. The second problem is that multiple regression assumes a particular pattern of random error in the data which is most improbable for dichotomous variables. Thus, multiple regression makes inappropriate assumptions for analyses where the dependent variable can take on only a few specific values. Indeed, the argument can be extended to show that multiple regression is inappropriate for any situation where the dependent variable is categorical, rather than continuous.

Loglinear models can step into the breach. The logistic models considered in this chapter do not suffer from the difficulties of multiple regression when applied to categorical dependent variables. Moreover, as we shall see, there is a clear analogy between the form of these models and multiple regression models, so that similar techniques can be applied to both. For this reason, the use of logistic models is often referred to as *logistic regression*.

Applied to the data of Table 10.1, logistic regression will allow us to develop and test models to predict the odds of voting Democrat rather than Republican. Putting it another way, given knowledge of a respondent's party identification, attitude and whether there is an incumbent up for election in the state, the model should predict the odds of that respondent voting Democrat rather than Republican. The form of the model needed to do this can be derived from the loglinear equation presented in the previous chapter. Suppose that there is a respondent with party identification *i* (where *i* could mean Democrat, Independent or Republican, depending on the respondent), attitude *j* (Democrat, Neutral or Republican) and Incumbency *k* (the respondent lives in a state where the incumbent is Democrat or Republican, or there is no incumbent). The odds of this respondent voting Democrat are:

$$m_{Dijk} / m_{Rijk}$$

This is the ratio of the counts in the cells for Democrats and Republicans, for the row of the table corresponding to party identification *i*, attitude *j* and incumbency *k*. The log odds, or *logit*, of the respondent voting Democrat are then:

$$\log\left(m_{Dijk} / m_{Rijk}\right) = \log m_{Dijk} - \log m_{Rijk}$$

From Chapter 9, we have the loglinear equation which expresses the log count as a sum of μ-terms. For a saturated model for the four variables, vote (V), party identification (P), attitude (A) and incumbency (I),

$$\begin{aligned}\log m_{lijk} &= \mu + \mu_l^V + u_i^P + \mu_j^A + \mu_k^I \\ &+ \mu_{li}^{VP} + \mu_{lj}^{VA} + \mu_{lk}^{VI} + \mu_{ij}^{PA} + \mu_{ik}^{PI} + \mu_{jk}^{AI} \\ &+ \mu_{lij}^{VPA} + \mu_{lik}^{VPI} + \mu_{ljk}^{VAI} + \mu_{ijk}^{PAI} + \mu_{lijk}^{VPAI}\end{aligned}$$

While this may look complicated, remember that each μ-term corresponds to one fitted marginal (the one indicated in the superscript of the μ-term) and that the equation merely represents the fact that, for a saturated model, all the possible marginals are fitted.

The log model table counts ($\log m_{Dijk}$ and $\log m_{Rijk}$) in the expression for the log odds can now be rewritten using the equation for $\log m_{lijk}$:

$$\begin{aligned}\log\left[\frac{m_{Dijk}}{m_{Rijk}}\right] &= \log m_{Dijk} - \log m_{Rijk} \\ &= \mu + \mu_S^V + \mu_i^P + \mu_j^A + \mu_k^I \\ &+ \mu_{Di}^{VP} + \mu_{Dj}^{VA} + \mu_{Dk}^{VI} + \mu_{ij}^{PA} + \mu_{ik}^{PI} + \mu_{jk}^{AI} \\ &+ \mu_{Dij}^{VPA} + \mu_{Dik}^{VPI} + \mu_{Djk}^{VAI} + \mu_{ijk}^{PAI} + \mu_{Dijk}^{VPAI} \\ &- \mu - \mu_R^V - \mu_i^P - \mu_j^A - \mu_k^I \\ &- \mu_{Ri}^{VP} - \mu_{Rj}^{VA} - \mu_{Rk}^{VI} - \mu_{ij}^{PA} - \mu_{ik}^{PI} - \mu_{jk}^{AI} \\ &- \mu_{Rij}^{VPA} - \mu_{Rik}^{VPI} - \mu_{Rjk}^{VAI} - \mu_{ijk}^{PAI} - \mu_{Rijk}^{VPAI}\end{aligned}$$

Fortunately, this lengthy equation can be greatly simplified because all the μ-terms not mentioning the dependent variable, V, cancel out. Furthermore, since $\mu_D^V = -\mu_R^V$ and similarly for other μ-terms including the V variable, all μ-terms mentioning Republican voters can be replaced by minus the corresponding μ-term for Democratic voters, leaving a much simpler expression:

$$\log\left[\frac{m_{Dijk}}{m_{Rijk}}\right] = 2(\mu_D^V + \mu_{Di}^{VP} + \mu_{Dj}^{VA} + \mu_{Dk}^{VI} + \mu_{Dij}^{VPA} + \mu_{Dik}^{VPI} + \mu_{Djk}^{VAI} + \mu_{Dijk}^{VPAI})$$

The result of this algebra is the basic logistic regression equation which shows that the log odds of voting Democrat can be predicted from twice the sum of a set of μ-terms, all involving the variable Vote.

If you are familiar with multiple regression models, you will know that the standard equation for multiple regression is something like:

$$y = \beta_0 + \beta_1 x_1 + \beta_2 x_2 + \beta_3 x_3$$

where y is the dependent variable and the x's are the independent variables. β_0 is a constant which does not vary with the values of the independent variables. The logistic regression equation above has much the same form. The first term on the right-hand side (μ_D^V) is also a constant, not varying with the values of the independent variables, P, A and I.

The above logistic equation represents the fitting of a saturated model to the data. It may be that a simpler model would fit adequately - for example, one in which there is no association or interaction effects, but only main effects. The equation for such a model is:

$$\log\left[\frac{m_{Dijk}}{m_{Rijk}}\right] = 2(\mu_D^V + \mu_{Di}^{VP} + \mu_{Dj}^{VA} + \mu_{Dk}^{VI})$$

The more complex terms have been omitted. This can be done because of the correspondence between loglinear effects and μ-terms. The question now is, does this simpler model fit the data?

Some computer programs for loglinear analysis (for example, SPSS) will fit logistic models directly. The dependent and independent variables and the form of the model are specified, and the program calculates the fit in terms of G-squared and the values of the effect parameters. Other programs do not have this facility, but it is easy to re-express a logistic regression as an ordinary loglinear model and then fit that loglinear model. The loglinear representation of a logistic model includes additional effects corresponding to all the associations and interactions between the *independent* variables. For example, the loglinear version of the main effects logistic model is:

$$V{*}P + V{*}A + V{*}I + P{*}A{*}I$$

This model includes each of the relationships between Vote and the independent variables, and also the interaction, $P{*}A{*}I$, which, because it includes the lower-order relatives, covers all the possible relationships between the three independent variables. Fitting this loglinear model will give exactly the same results (the same G-squared, degrees of freedom and effect parameters) as fitting the logistic model directly.

Fitting the main-effects logistic model gives a G-squared of 11.8 with 20 degrees of freedom and a significance level of 92 per cent. This model fits very well, showing that there are no significant interactions between Vote and pairs of the other variables. Next, each of the three models obtained by deleting one of the associations between Vote and the independent variables can be tested. None of them fits, suggesting that all these associations are required. The main-effects logistic model is the simplest that fits the data.

Table 10.2 Effect parameters for a main effects logistic model fitted to the data of Table 10.1.

(a) Vote by party identification	Vote	
Party identification	Democrat	Republican
Democrat	0.76	−0.76
Independent	0.00	0.00
Republican	−0.76	0.76
(b) Vote by attitude		
Attitude	Democrat	Republican
Democrat	0.30	−0.30
Neutral	0.08	−0.08
Republican	−0.38	−0.38
(c) Vote by incumbency		
Incumbency	Democrat	Republican
Democrat	0.20	−0.20
No	−0.01	0.01
Republican	−0.19	0.19

Table 10.2 shows the values of the μ-terms for each of the associations between Vote and the other variables that were calculated when the main-effects model was fitted. These values can be used to draw conclusions about the relationship between voting Democrat and the explanatory variables. For example, what is the effect on Democratic voters of having an incumbent Democrat? This question can be answered by finding an odds ratio: the ratio of the odds of voting

Democrat when a Democrat is incumbent and the odds of voting Democrat when there is no incumbent. The odds of voting Democrat when there is a Democrat incumbent are:

$$\log\left[\frac{m_{DijD}}{m_{RijD}}\right] = 2(\mu_D^V + \mu_{Di}^{VP} + \mu_{Dj}^{VA} + \mu_{DD}^{VI})$$

and the odds of voting Democrat when there is no incumbent are:

$$\log\left[\frac{m_{DijN}}{m_{RijN}}\right] = 2(\mu_D^V + \mu_{Di}^{VP} + \mu_{Dj}^{VA} + \mu_{DN}^{VI})$$

The log odds ratio is then obtained by subtracting the two right-hand sides:

$$\log\left[\frac{m_{DijD}}{m_{RijD}} \Big/ \frac{m_{DijN}}{m_{RijN}}\right] = \log\left[\frac{m_{DijD}}{m_{RijD}}\right] - \left[\frac{m_{DijN}}{m_{RijN}}\right]$$

$$= 2(\mu_{DD}^{VI} - \mu_{DN}^{VI})$$

Table 10.2(c) gives the values of μ_{DD}^{VI} and μ_{DN}^{VI} as 0.20 and −0.01, so the log odds ratio is $2 \times [0.20 - (-0.01)] = 0.42$ and the odds ratio itself is 1.23, meaning that respondents are 1.23 times more likely to vote Democrat if there is a Democratic incumbent than if there is no incumbent standing again for election. The log odds ratio for voting Democrat when there is a Republican incumbent compared with no incumbent is $2 \times (-0.19 - 0.01) = -0.40$, and the odds ratio is 0.67, so respondents are less likely to vote Democrat when there is a Republican incumbent, other things being equal.

An important point to remember about these conclusions is that fitting the main-effects logistic model allows us to say that the effects of incumbency on voting hold regardless of respondents' party identification and attitude. Because we have shown that there are no association and interaction effects, the odds ratios apply to all the respondents, independently of their values on the other explanatory variables.

Another example of the usefulness of logistic models can be found in an analysis of the receipt of caring services by elderly infirm people living at home. Using data from the 1980 General Household Survey, Arber & Gilbert (1989) examined the common assumption that, because of their gender, elderly disabled men obtain much more support from the statutory social services and voluntary organizations than women do. Simply looking at the quantity of such services that elderly men and women receive can be misleading, because the need for services depends for obvious reasons on the elderly persons' level of disability and whether they live alone or with other people who can give them support. A higher proportion of elderly women are severely disabled, and more elderly women than men live in single-person households.

A better procedure is to model the receipt of services as the dependent variable with level of disability and the characteristics of the household in which the elderly person lives as explanatory variables, using logistic regression. Three forms of community support were considered: the supply of home helps by the local authority; Meals-on-Wheels, a voluntary service that brings hot meals to elderly people in their homes; and visits from the district nurse.

Table 10.3 shows the results of a logistic regression of the odds of receiving each of these forms of support for men and women living in different types of household. The figures were obtained by fitting a main-effects logistic model with type of household and level of disability as the independent variables, separately for each of the three forms of support. These models fitted well at the 5 per cent level. Because odds ratios compare the odds of being in one category with the odds of being in another, it is helpful to select one category as the "standard" and make all comparisons with this category. In Table 10.3, the likelihood of an elderly married couple receiving a service has been selected as the standard against which the likelihood of receipt of services by people living in other types of household is compared. The table shows that, after controlling for levels of disability, elderly men and women who live alone are over five times more likely to receive home help support than elderly married couples. Men living alone are 5.66/5.21, about 8 per cent, more likely to have a home help than women living on their own and about 14 per cent more likely to have been visited by a district nurse in the last month, but are about 300 per cent more likely to have received Meals-on-Wheels. Thus, although once disability level has been controlled there is a clear difference between men and women living alone in the likelihood of their getting Meals-on-Wheels from the voluntary sector, differences in the likelihood of receipt of public-sector services, such as home helps and district nursing, are relatively small and are outweighed by the differences between types of household.

Table 10.3 Receipt of services by elderly people living in different types of household.

Type of household including an elderly person	Likelihood§ of receipt of: Home helps	Meals on wheels	District nursing
Living *alone*			
Man	5.66	15.27	1.91
Woman	5.21	5.22	1.68
Living with *spouse*			
Couple, both elderly	1.00	1.00	1.00
Couple, one elderly, one younger	0.24	0.00	0.56
Couple and adult children	0.00	0.00	0.90
Living with *siblings* or other elderly	1.29	2.69	2.84
Living with *unmarried*			
male adult child	0.74	1.48	1.43
female adult child	0.80	1.37	1.28
Living with *married* child (or with child who is a lone parent)	0.30	0.36	0.72

§ Odds ratio of receipt of the service within the last month, controlling for level of disability.
Source: Arber & Gilbert (1989: 115).

Dependent variables with several categories

In the examples so far, the dependent variable has had just two categories. There is a version of logistic regression that uses the *multinomial logit* model, which can handle dependent variables with more than two categories, although it is often then more difficult to arrive at clear and sociologically meaningful interpretations of the resulting odds ratios. Table 10.4 cross-tabulates data from the 1988 General Household Survey on the economic status of married women aged 50–65. These data can be used to examine some of the factors that influence older women's labour force participation. The dependent variable has categories for women working full-time (more than 30 hours a week), part-time and not in paid work. The explanatory variables are the woman's age, her health status (whether she has a long-standing illness which limits her normal activities), and whether her husband is in paid work. Forward selection shows that the simplest logit model that fits these data includes the two interaction effects: the interaction between economic status, husband's economic status and age, and the interaction between economic status, husband's economic status and health. This suggests that both the relationships between a woman's age and her economic status and between her health and her economic status are affected by whether her husband works.

Table 10.4 Wife's economic status by age, whether suffering from a limiting long-standing illness, and husband's economic status.

Husband in paid work:	Yes						No					
Long-standing illness:	Yes			No			Yes			No		
Age:	50–54	55–59	60–64	50–54	55–59	60–64	50–54	55–59	60–64	50–54	55–59	60–64
Full-time work	189	171	115	427	276	156	5	3	4	19	16	4
Part-time work	59	62	35	120	93	38	9	19	9	15	29	18
Not in paid work	51	72	50	80	49	59	86	147	347	45	101	187

Source: OPCS (1990); data extracted by S. Arber.

Further detail on this can be obtained by examining the effect parameters. Table 10.5 shows odds ratios corresponding to the associations and interactions in this model, referred to the situation of a woman aged 50–54 in full-time employment with a working husband and no longstanding illness. The odds ratios given as dots (•) are fixed to be unity by the choice of a standard situation for reference. Several of the odds ratios are not significantly different from unity at the 5 per cent level and these are shown with a *1* in italics.

The odds of not being in paid work rather than working full-time are over 14 times greater for those women whose husbands are not in work than for those whose husbands are working. Thus, there is a strong tendency for the wife not to have a paid job if her husband does not. Also, if her husband is not in work, a wife is nearly three times as likely to be in part-time work as full-time work, compared with couples where the husband is working. The likelihood that a woman does not have a paid job is increased by a factor of 1.8 if she is aged over 60 and by 1.6 if she has a limiting long-standing illness, but these variables have no significant effect on part-time working.

Table 10.5 Odds ratios from fitting a model to the data of Table 10.4.

Model fitted: Interactions between wife's economic status, husband's economic status and wife's age, and wife's economic status, husband's economic status and wife's health status, and all their lower-order relatives. G^2 = 11.0 d.f. = 8. Significance = 20%

(a) Wife's economic status by husband's economic status

	Husband: Working	Not in work
Full-time work	•	•
Part-time work	•	2.9
No paid work	•	14.3

(b) Wife's economic status, by wife's health

	Longstanding illness: Does not have illness	Has illness
Full-time work	•	•
Part-time work	•	*1*
No paid work	•	1.6

(c) Wife's economic status by wife's age

	Age: 50–54	55–59	60–64
Full-time work	•	•	•
Part-time work	•	1	1
No paid work	•	*1*	1.8

(d) Wife's economic status by husband's economic status by wife's age

	Husband: Working	Not in work
Age 50–54		
Full-time work	•	•
Part-time work	•	•
No paid work	•	•
Age 55–59		
Full-time work	•	•
Part-time work	•	2.2
No paid work	•	2.1
Age 60–64		
Full-time work	•	•
Part-time work	•	3.7
No paid work	•	6.8

(e) Wife's economic status by husband's economic status, by wife's health

	Husband: Working	Not in work
Does not have longstanding illness		
Full-time work	•	•
Part-time work	•	•
No paid work	•	•
Has longstanding illness		
Full-time work	•	•
Part-time work	•	*1*
No paid work	•	3.5

Key: • unity because of contrasts. *1* estimate not significantly different from unity at the 5% level.

The interaction effect of economic status by husband's status by age shows that, comparing wives whose husbands are not working with those whose husbands are working, there is an increasing tendency to leave the labour market with age, so that, for example, the odds of not having a paid job (compared with the odds of working full-time) are nearly seven times greater for women who are aged over 60 and whose husbands are not working than the odds would be considering either of these factors separately. Finally, the interaction between economic status, husband's economic status and health is most marked amongst women who have a long-standing limiting illness and whose husbands are not working. These women are 3.5 times less likely to work full-time than would be expected from considering either their health or their husband's economic status separately.

These odds ratios can be combined to compare women in particular circumstances. For example, a woman aged between 55 and 59 who has a long-standing illness and an unemployed husband is

$$14.3 \times 1.6 \times 1 \times 2.1 \times 3.5 = 168$$

times less likely to be in full-time work compared with the reference category of women aged 50–54 in good health and with husbands in work. The numbers in this product can be read from Table 10.5, taking one number from each subtable according to the categories being compared.

The example has shown that logit models for response variables with more than two categories can be fitted quite straightforwardly using exactly the same methods as models for two-category response variables. The complexity comes mainly in interpreting the results. This is especially the case where there are not only main effects but also interactions to deal with.

Summary

A type of loglinear model, the logit, may be used to examine how a categorical dependent or response variable is related to one or more explanatory or independent variables. Logistic regression is concerned with modelling the odds of this dependent variable, rather than the cell counts, which are what loglinear models aim to predict. Logistic regression has similarities to multiple regression, a method used when the dependent variable is continuous and measured at the interval level. The parameters for logistic regression are most easily interpreted if they are expressed as odds ratios.

Computer analysis

The following examples show how logistic regression can be performed using SPSS and GLIM. Both offer special procedures for fitting logistic models, although the standard loglinear methods can also be used if the logistic model is transformed to the equivalent loglinear model.

SPSS

Program 10.1 fits a main-effects logistic model to the data of Table 10.1. (Some lines of data have been omitted to save space.) For logistic regression, SPSS requires that the dependent variable is listed first after the LOGLINEAR command, followed by the keyword, BY, and the independent variables. As usual with the LOGLINEAR command, the DESIGN subcommand specifies the model to be fitted, and all lower-order relatives must be included in the model specification. It is therefore necessary to include in the DESIGN subcommand not just the associations between Vote and the other variables, but also the main effect of Vote itself.

Program 10.1

```
DATA LIST FREE/ PartyID, Attitude, Incumb, Vote, Count.
BEGIN DATA
1    1    1    1    288
1    1    1    2    25
1    1    2    1    139
1    1    2    2    18
................
3    3    1    2    155
3    3    2    1    13
3    3    2    2    129
3    3    3    1    14
3    3    3    2    227
END DATA.
WEIGHT BY Count.
LOGLINEAR Vote(1,2) BY PartyID(1,3), Attitude(1,3),
   Incumb(1,3)
 /PRINT= ESTIM
 /DESIGN= Vote, Vote BY PartyID, Vote BY Attitude,
          Vote BY Incumb.
```

The output from this program includes the effect parameters (the μ-terms) for each effect specified in the DESIGN subcommand. The values provided are those shown in Table 10.2.

GLIM

GLIM's procedure for fitting logistic models is intended for modelling *proportions*, not the odds we have been considering so far. The difference is that a proportion is the number falling into a category divided by the marginal (e.g. 288 Democrats as a proportion of the total of 313 voters), while the odds are the number in one category divided by the number in another category (e.g. the odds of 288 Democrats against 25 Republicans). For tables such as the voting data of Table 10.1, expressing the response variable in terms of Democratic voters as a proportion of all voters only involves dividing each cell count by the count from the Vote marginal. The results of the analysis are exactly equivalent to one based on odds.

An important difference that follows from modelling proportions rather than odds is the nature of the sampling distribution which should be assumed by the analysis. Because the data come from a random sample, the count in a particular cell of a table is always likely to differ from the true one by some random amount. If many samples are drawn and counts are made from each sample, the distribution of these counts is called the *sampling distribution*.

For cross-tabulations, the sampling distribution of each cell count is usually taken to follow a *Poisson* distribution, which is appropriate when counting events that occur randomly with outcomes that are independent (e.g. a respondent's vote does not depend in any way on other respondents' vote). It is because the Poisson distribution is appropriate for counts that previous examples of fitting loglinear models using GLIM have included the directive "$error p", which specifies that the error structure to be used is Poisson.

However, when modelling proportions, there is an additional point that must be taken into account: a proportion cannot exceed 1 in magnitude. In this situation, the appropriate sampling distribution is the *Binomial* distribution. This is indicated to GLIM using the "$error B" directive, which is followed by the denominator of the proportion (e.g. "$error B N", where N is the number of voters).

In Program 10.2, the numbers of Democrat and Republican voters for each combination of party identification, attitude and incumbency are read in. The total number of voters for each combination is calculated and stored in the variable N with the first $calc directive. The %GL (generate levels) function is used to construct the values of the three independent variables (factors). The dependent variable, the number of Democratic voters, is defined with the $yvar directive and the error structure is declared to be binomial. Finally, the model to be fitted is specified as a main-effects logistic for the three independent variables.

Program 10.2

```
$units 27
$data Dem Rep
$read
288  25 139  18 193  37 185  18  83
 15  99  31  49  20  29  14  38  29
 48  13  14   6  27  21  81  51  20
 29  50  70  30  51  18  39  21  63
  7  19   2   8   4  28  20  63  13
 49  19  88  23 155  13 129  14 227
$calc N = Dem + Rep
$factor PartyId 3 Attitude 3 Incumb 3
$calc PartyId= %GL(3,9) : Attitude= %GL(3,3) : Incumb= %GL(3,1)
$yvar Dem
$error B  N
$fit PartyId + Attitude + Incumb
$display E $
```

The output from this program is shown in Table 10.5. The parameter estimates are the log odds ratios with the first category of each effect treated as the standard for comparison and its value therefore set to zero. For example, the log odds ratio of voting Democrat with a Democrat incumbent compared with voting Democrat with no incumbent is equal to the estimate for the first category of INCU (Democratic incumbent), minus the estimate for the second category (No incumbent), or $0 - (-0.42) = 0.42$

The parameter values in Table 10.5 are all more than 1.96 times their standard errors (s.e.), indicating that they are all statistically significant.

Table 10.5 GLIM parameter estimates from a main effects logistic regression of the data in Table 10.1.

	Estimate	s.e.	Parameter
1	2.474	0.1233	1
2	−1.537	0.1166	PART(2)
3	−3.068	0.1338	PART(3)
4	−0.4338	0.1260	ATTI(2)
5	−1.371	0.1348	ATTI(3)
6	−0.4200	0.1320	INCU(2)
7	−0.7948	0.1128	INCU(3)

Further reading

There are many sources for more details on logistic models, which have become very popular among statisticians because of their similarity with normal regression models for interval level variables. Hosmer & Leleshow (1989) provide an exhaustive treatment. Agresti (1990) introduces logistic models in the context of loglinear modelling. Aitkin et al. (1989, Chs 4 and 5) discuss logistic models and relate them to other regression models, using GLIM for their examples. Alba (1988) explains how to interpret the parameters of complex logistic models in terms of odds ratios.

CHAPTER ELEVEN
Causal analysis

Discussing the interpretation of the results of loglinear analysis in previous chapters, we have occasionally made assumptions about the causal ordering of variables. For example, analyzing the data in Table 4.1, we found an interaction between gender, income and occupation. We assumed that this relationship was due to the causal influence of gender on the association between income and occupation. The result of the statistical analysis (that there is an interaction) does not itself provide evidence to prove the existence or the direction of causality. Causal statements may be justified using common knowledge of the world, the temporal ordering of events, the results of experiments, or through theorizing, but they cannot be proved solely from statistics.

The starting-point for a causal analysis must therefore be some assumptions about the causal ordering of the variables. A convenient way of summarizing and presenting these assumptions is a *causal diagram*. This is a diagram in which arrows point from the assumed causal variables to the assumed affected variables. An example based on the model used in Blau & Duncan's (1967) research on the American male occupational structure is shown in Figure 11.1. In this diagram the antecedent, causally prior variables are on the left and those that are affected by these causes are on the right, so the arrows tend to go from left to right. The curved, double-headed arrow in the figure indicates an assumed association between father's education and status for which no particular statement about the direction of causality is being made. The variables to the left of the diagram are the explanatory variables, because they can be used to explain the values taken by the variables on the right. Thus, the model assumes that, given a father's score on the status and education variables, it should be possible to predict his son's educational score. The father variables must be causally prior to the respondent variables, and respondent's education must be prior to the respondent's present status.

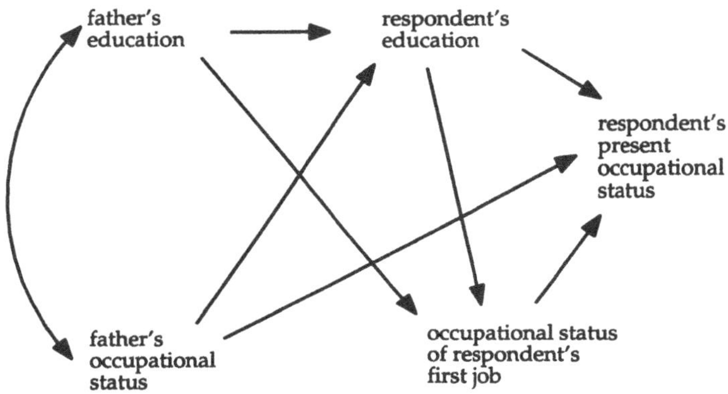

Figure 11.1 Causal diagram, showing hypothesized causal relationships between fathers' and sons' education and occupational status (derived from Blau & Duncan 1967: 70).

Ordinary ideas of causality require that causes must come before their effects. Therefore, if there is a temporal order among the variables, the causal model must be chosen to follow this order. Because the variables in Figure 11.1 do have a clear order in time, sorting out which variables are explanatory and which are consequences is easy.

In other cases, making sensible assumptions about causal ordering is very much more difficult. The difficulties may arise because it is not clear which variables are causes and which effects. Or the difficulties may stem from the feeling that the causation is not unidirectional: *A* causes *B*, but *B* also causes *A*. Another complication is the treatment of interaction between three or more variables, which cannot easily be represented by a single arrow in a diagram. As noted in Chapter 4, an interaction between variables *A*, *B* and *C* may be described in any of three equivalent ways: as the association between *A* and *B* varying according to the levels of *C*; as the association between *B* and *C* varying according to the levels of *A*; or as the association between *A* and *C* varying according to the levels of *B*. In general, it is best to select the causally prior variable to be the one that affects the association between the other two, but often it is not easy to determine which of the three is in fact causally prior.

Putting these difficulties to one side, we shall now return to the data on US senatorial elections introduced in the previous chapter for an example of causal analysis. The first necessity is to determine some causal ordering between the four variables, Vote, Party identification, Attitude and Incumbency. There is clearly room for debate about an appropriate causal sequence. Nevertheless, we shall take Party identification and Incumbency as joint explanatory variables, hypothesizing some relationship between them, but making no commitment about its direction. Party identification is taken as a cause on the ground that this is a fairly stable feature of people's political values, one that would usually have been established long before voting. We shall assume that these two variables causally influence both Attitude to the parties' policies on domestic issues and management ability, and Voting. Moreover, we shall assume that Attitude is causally prior to Voting. The last assumption is in accord with common sense – what one thinks of the parties' policies and competence affects one's vote – but common sense is not an infallible guide. It could be argued, for instance, that these attitudes were measured in interviews some time after the votes had been cast, and the way one votes might affect how one subsequently judges the competence of the parties.

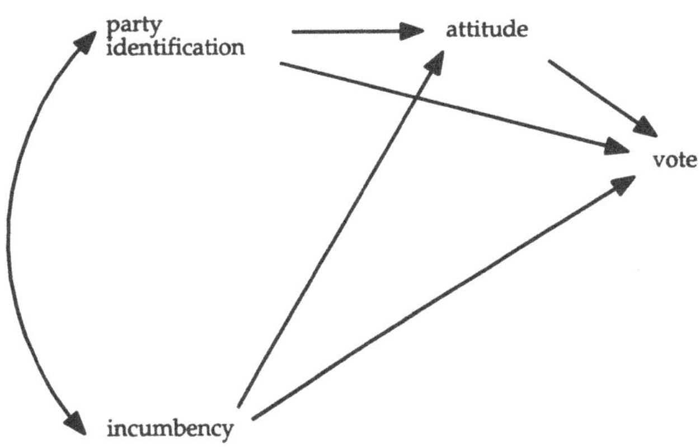

Figure 11.2 Causal model, showing all possible two-variable paths for US senatorial elections data.

We shall, however, stick with the assumption that Attitude influences Voting, relying on the argument that, although the attitudes were indeed measured after the votes were cast, they are unlikely to have shifted sharply from those held before the elections took place. These considerations lead to the causal model diagrammed in Figure 11.2, which shows all the possible two-variable causal relationships that could exist, given the causal sequence we have settled on. There may also be interactions between the variables, but including these too would have made the diagram difficult to follow. The next task will be to pare down to the minimum the number of causal influences or "paths" shown in the diagram by eliminating those that are not needed to fit the data. This will involve the examination of a number of separate logistic models, one after the other.

Starting at the left-hand side of the diagram, we can ask whether there is actually a significant association between the two predictor variables - in other words, is the curved arrow really necessary? This question can be answered by examining data relating only to the two variables, Party identification and Incumbency, ignoring the other variables because we have assumed that they exert no causal influence on these predictors. To check whether there is a statistically significant association between Party identification and Incumbency, a model of no association can be fitted to the marginal table of Party identification and Incumbency derived from Table 10.1.

The results in Table 11.1 indicate that the no-association model (model 1) fails to fit the data, implying that there is association between Party identification and Incumbency and that the curved arrow between these variables in Figure 11.2 should be retained. Next, to investigate the causal paths to the Attitude variable, a main-effects logistic model with this as the dependent variable and Party identification and Incumbency as the independent variables (model 2) is fitted to the marginal table of Attitude, Party identification and Incumbency. The fit is good, suggesting that there is no interaction between the three variables. Variations on this model, omitting each of the terms in turn, show that the fit remains good when the Incumbency effect (i.e. the marginal $A*I$) is omitted, but becomes poor if the Party identification effect ($A*P$) is left out. This means that there is no significant relationship between Attitude and Incumbency and that the causal arrow in Figure 11.2 representing this relationship can be deleted, but the other arrow pointing to Attitude must be retained. Finally, a logistic model with Vote as the dependent variable and the others as the independent variables (model 3) is examined (this is the same model as that considered in the previous chapter), using the full data table. As Table 11.1 shows, none of the independent variables may be omitted from the model without reducing the fit unacceptably.

Table 11.1 Results of fitting models to the data of Table 10.1.

	Model	Table	Fitted marginals	G^2	df	Significance
(1)	No association between Party identification and Incumbency	*P* by *I*	$P+I$	41	4	0%
	Logistic model with Attitude as dependent variable and Party identification and Incumbency as independent variables:	*P* by *I* by *A*				
(2)	All main effects		$A*P + A*I + P*I$	10	8	28%
(2.1)	less $A*I$		$A*P + P*I$	18	12	11%
(2.2)	less $A*P$		$A*I + P*I$	846	12	0%
	Logistic model with Vote as the dependent variable and Party identification, Incumbency and Attitude as independent variables:	*P* by *I* by *A* by *V*				
(3)	All main effects		$V*P+V*I+V*A+P*A*I$	12	20	92%
(3.1)	less $V*P$		$V*I+V*A+P*A*I$	676	22	0%
(3.2)	less $V*I$		$V*P+V*A+P*A*I$	62	22	0%
(3.3)	less $V*A$		$V*P+V*I+P*A*I$	126	22	0%

The conclusion of this investigation is that only the causal link between Incumbency and Attitude may be omitted, meaning that there is evidence for the influence of Attitude, Party identification and Incumbency on Voting behaviour, and that Attitude is influenced by Party identification, but not by Incumbency (see Figure 11.3).

As a second illustration of this method of testing causal models, we shall examine further the data on extra- and premarital sex that we first encountered in Chapter 3 (see also Programs 8.1 and 8.2). The four variables that were tabulated in Table 3.3 are Gender (*G*), whether the respondent reported experience of pre-marital sex (*P*), whether the respondent reported experience of extramarital sex (*E*), and the respondent's marital status at the time of the survey (*M*). These variables can be arranged in the temporal sequence, *G, P, E, M*. Gender is first in time order because it cannot be affected by any of the other three variables. The incidence of Premarital sex may be affected by Gender, but not by the other two variables. Extramarital sex may be explained by Pre-marital sexual experience and Gender, but not by current Marital status. And Marital status may be influenced by any or all of the other three variables.

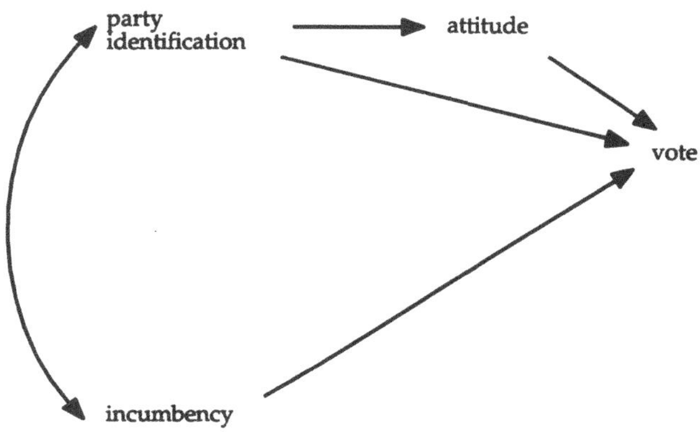

Figure 11.3 Causal model, showing causal paths for US senatorial elections data.

One possible causal model which conforms with this temporal sequence is shown in Figure 11.4 (Agresti 1990). The figure indicates only potential direct relationships between pairs of the variables; there may also be interaction effects.

To investigate this model, we begin at the left-hand side of the diagram, with the hypothesized effect of Gender on Pre-marital sexual experience. This path can be tested by fitting a simple model of no association between *G* and *P* to the marginal table of Gender by Premarital sex.

It is often more convenient, however, to test for the association on the basis of the full, four-dimensional table (Table 11.2). This saves calculating the marginal table. The association can be tested by comparing the fit of two models: one that just includes the association effect, and one that has only the main effects of the two variables. The differences in G-squared and degrees of freedom for the two models are a test for the presence of the association. The difference in *G*-squared is 75 with 1 degree of freedom, indicating that the independence model

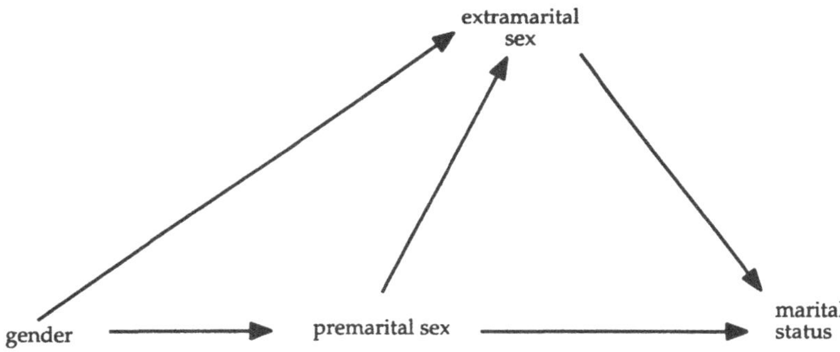

Figure 11.4 Causal diagram relating gender, reported experience of pre- and extramarital sex, and marital status.

does not fit and confirming the link between Gender and Premarital sex.

Next, we consider the potential causal influences on Extramarital sex: Gender and Premarital sex. First, we must test whether there is an interaction between these three variables. This is done by contrasting a model including the interaction with a model without the interaction but with all its lower-order relatives. The difference in G-squared is close to zero, confirming the absence of interaction. To examine whether the path from P to E is needed, we contrast a model including the effect $P*E$ with one without this effect. Both of these models should include the $G*P$ effect which as we have already seen is required. The difference in G-squared is 46 with 1 degree of freedom, indicating that this path is needed. In a similar way, we can test the path between G and E, basing the comparisons on models which include both $G*P$ and $P*E$. The difference in G-squared is only 3, with 1 degree of freedom, suggesting that this path is not statistically significant.

Parameter estimates from fitting the model

$$G*P + P*E$$

show that the log of the ratio of the odds of reporting extra-marital sex for those reporting premarital sex is 1.38 (equivalent to an odds ratio of 4.0). Thus, the estimated odds of extramarital sex are four times higher for those who reported premarital sex than for those who did not. The fact that the path from Gender is not required means that all the effect of gender on the likelihood of reporting extramarital sex is explained by

Table 11.2 Results of fitting models to the data of Table 3.3.

	Fitted marginals	G^2	d.f.	Significance
	$G+P$	923	13	
	$G*P$	848	12	
Path between G and P		75	1	0%
	$G*P+G*E+E*P$	110	9	
	$G*P*E$	110	8	
Interaction between G, P and E		0	1	97%
	$G*P+E$	159	11	
	$G*P+E*P$	113	10	
Path between P and E		46	1	0%
	$G*P+E*P$	113.1	10	
	$G*P+E*P+E*G$	110.2	9	
Path between G and E		2.9	1	8%
	$G*P*E+G*M+P*M+E*M$	13.6	4	
	$G*P*E+G*P*M+G*E*M+P*E*M$	0.1	1	
Interaction between P, E and M		13.5	3	0%
	$G*P*E+G*M+P*E*M$	0.698	3	
	$G*P*E+G*P*M+G*E*M+P*E*M$	0.146	1	
Interactions between G, E and M and between G, P and M		0.552	2	76%
	$G*P*E+P*E*M$	5.25	4	
	$G*P*E+G*M+P*E*M$	0.70	3	
Path between G and M		4.55	1	3%

the different propensity of men and women to report premarital sex.

Finally, we can test for the presence of interaction between Marital status and pairs of the other variables, by comparing the fit of a model including these interactions with one including effects for all the possible causal paths to Marital status. The difference in G-squared is 13.5 with 3 degrees of freedom, indicating that there is interaction present. The site of this interaction can be narrowed down by replacing the effects involving interactions of Gender and Marital status with Pre-

and Extra-marital sex (G^*P^*M and G^*E^*M) with the association from Gender to Marital status. The comparison of this model with the previous one suggests that neither of these interactions is present. A further comparison with a model omitting the path from Gender to Marital status shows that, using the 5 per cent criterion, this path is not required either.

This last model predicts that the effects of extra-marital sex on marital status (i.e. on whether there has been a divorce) depends on whether the respondent has reported pre-marital sex. The parameter estimates show that the odds of divorce when extra-marital sex has been reported are increased by a factor of 6 when pre-marital sex is not also admitted. The consequences of extra-marital sex on a marriage may be much more dramatic for couples where there has been no experience of sexual relations with others before the marriage.

Several points need to be made about the procedure we have followed to examine the causal paths for these data. First, conclusions about the need for paths depended on comparing two models, one including the effect of interest and one excluding it. The *G*-squared difference obtained from such comparisons between models is distributed as χ^2, provided that one model is *nested* in the other. Nesting means that it must be possible to obtain the simpler model by deleting effects from the more complex one. No effects may be added. For example, the model G^*P^*E, when expanded to mention explicitly all the lower-order relative effects, becomes:

$$G.P.E + G.P + G.E + P.E + G + P + E$$

The model $G^*P + G^*E + E^*P$ is nested within G^*P^*E because it can be obtained by deleting the effect $G.P.E$.

Secondly, the models that were chosen for comparison were selected so that:

(a) effects representing paths which were causally prior to the ones being examined were *included* in the model, and
(b) effects representing paths which followed in causal order the paths being examined were *omitted* from the model.

For example, to study the path between Pre-marital Sex and Extra-marital sex, the effect G^*P, representing the path between Gender and Pre-marital sex, was included and all effects representing paths to the Marital status variable were excluded from the model (see Table 11.2).

Thirdly, although we have described the models as loglinear models, they can all be reformulated as logistic models. Recall that in Chapter 10 it was stated that a loglinear model can be derived from a logistic model by adding all the interactions and associations between the independent variables. For example, a logistic model with A as the response variable and B and C as explanatory variables is equivalent to the loglinear model

$$A*B + A*C + B*C$$

The comparison we made in order to test the path between Pre- and Extra-marital sex could have been obtained by testing two logistic models, both with Extra-marital sex as the response variable: one with Pre-marital sex as the explanatory variable and the other with no explanatory variables.

Summary

Once a diagram of causal relationships between variables has been formulated, it is possible to use loglinear or logistic models to test the presence of each of the hypothesized causal paths. The variables proposed as causes are treated as explanatory variables and the variables affected by these causes are response variables. The models can be used to identify paths where the relationship is not significant and which are therefore unnecessary, and to estimate odds ratios to assess the impact of the causal variables.

Computer analysis

The programs below examine the causal model of Figure 11.4 using the data of Table 3.3. They take slightly different approaches. The program using SPSS, Program 11.1, uses logistic regressions, while Program 11.2, for GLIM, fits equivalent loglinear models.

SPSS

Program 11.1 first reads the data and labels the variables. Three separate logistic fits are performed, the first two on marginal tables of the complete data table. The /PRINT=NONE causes SPSS to print neither the fitted counts nor the parameter estimates. The only results printed are the goodness-of-fit statistics, which are all that are needed for an initial exploration of the causal model.

In specifying logistic models for the LOGLINEAR command, SPSS

Program 11.1

```
DATA LIST FREE / MarStat, EMS, PMS, Gender, Count.
BEGIN DATA.
1 1 1 1  17
1 2 1 1  54
1 1 2 1  36
1 2 2 1 214
1 1 1 2  28
1 2 1 2  60
1 1 2 2  17
1 2 2 2  68
2 1 1 1   4
2 2 1 1  25
2 1 2 1   4
2 2 2 1 322
2 1 1 2  11
2 2 1 2  42
2 1 2 2   4
2 2 2 2 130
END DATA.
VARIABLE LABELS
 MarStat 'Marital Status'
 EMS 'Extra-marital Sex'
 PMS 'Pre-marital Sex'.
VALUE LABELS
 MarStat 1 'Divorced' 2 'Married'/
 EMS, PMS 1 'Yes' 2 'No'/
 Gender 1 'Women' 2 'Men'.
WEIGHT BY Count.
LOGLINEAR PMS (1,2) BY Gender(1,2)
 /PRINT= NONE
 /DESIGN= PMS, Gender.
LOGLINEAR EMS(1,2) BY Gender(1,2), PMS(1,2)
 /PRINT= NONE
 /DESIGN= EMS,EMS BY Gender, EMS BY PMS
 /DESIGN= EMS
 /DESIGN= EMS, EMS BY PMS.
LOGLINEAR MarStat(1,2) BY Gender(1,2), PMS(1,2), EMS(1,2)
 /PRINT= NONE
 /DESIGN= MarStat, MarStat BY Gender, MarStat BY PMS,
    MarStat BY EMS
 /DESIGN= MarStat, MarStat BY Gender, MarStat BY PMS,
    MarStat BY EMS,
    MarStat BY Gender BY PMS,
    MarStat BY Gender BY EMS,
    MarStat BY PMS BY EMS
 /DESIGN= MarStat, MarStat BY Gender, MarStat BY PMS,
    MarStat BY EMS,
    MarStat BY PMS BY EMS
 /DESIGN= MarStat, MarStat BY PMS, MarStat BY EMS,
    MarStat BY PMS BY EMS.
```

requires that all the effects to be included are listed in the DESIGN subcommand, including all lower-order relatives. The results from the program are sufficient to obtain all the figures shown in Table 11.2.

GLIM

The third line of Program 11.2 uses the %GL function to create the factor levels for each of the four variables, which saves having to type them in (compare the SPSS program, Program 11.1).

Each line beginning $fit first specifies a model to be fitted and then the same model with one or more additional effects. When GLIM displays the goodness-of-fit statistics for the augmented model, it also shows the changes in G-squared and degrees of freedom resulting from the additional effects. This is convenient for assessing the significance of effects by model comparison. The sequence of fits shown will exactly reproduce the sequence of models listed in Table 11.2.

Program 11.2

```
$units 16
$factor M 2 E 2 P 2 G 2
$calc M = %GL(2,8) : E = %GL(2,1) : P = %GL(2,2) : G =
%GL(2,4)
$data Count
$read
  17  54  36 214
  28  60  17  68
   4  25   4 322
  11  42   4 130
$yvar Count
$error p
$fit G + P                       $fit + G*P    $
$fit G*P + G*E + E*P             $fit + G*P*E  $
$fit G*P + E                     $fit + E*P    $
$fit G*P + E*P                   $fit + E*G    $
$fit G*P*E + G*M + P*M + E*M    $fit + G*P*M + G*E*M + P*E*M $
$fit G*P*E + G*M + P*E*M         $fit + G*P*M + G*E*M  $
$fit G*P*E + P*E*M               $fit + G*M    $
```

Further reading

Two articles by Goodman (1973a,b) laid the foundations for this approach to causal analysis in sociology. Fienberg (1980, Ch. 7) deals with the topic in detail. Causal models, graphical representations of them and some methods for causal analysis are discussed in Blalock (1964). Techniques for causal analysis based on multiple regression of interval variables are described in Heise (1975) and Duncan (1975).

CHAPTER 12
Models with ordinal variables

While many sociological variables are categorical, having no intrinsic ordering among their categories, there are also very many that are ordinal. For example, attitude scales with values ranging from "Agree strongly" to "Disagree strongly" are generally measured at the ordinal level. So are income groupings, measures of severity of illness, educational achievement and a host of other variables. As we noted in Chapter 2, categories can be compared in ordinal scales (for example, a response of "Very satisfied" indicates more satisfaction than a response of "Fairly satisfied"), but the distance between the categories cannot be determined ("Fairly satisfied" does not involve half as much, or any other measurable amount less, satisfaction than "Very satisfied"). So far, we have taken no notice of whether the relationships we have been modelling have been based on variables measured using a categorical or an ordinal scale. We have treated all variables as though they were categorical. In this chapter, we introduce models that take advantage of the extra information available in ordinal measurements.

Table 12.1 Opinion about tax rate for high income earners by socio-economic group.

Opinion on tax rate for high earners	Socio-economic group						
	Professional	Employer	Non-manual Int.	Non-manual Jun.	Skilled	Semi-skilled	Unskilled
Much too low	5	8	13	12	33	23	12
Too low	14	73	67	99	154	119	32
About right	28	106	78	164	145	115	26
Too high	20	62	40	58	62	43	10
Much too high	5	16	12	26	16	14	3

Source: SCPR (1987).

Table 12.1 shows respondents' opinions about whether tax rates for those with high incomes are too high, about right or too low, cross-classified with their socio-economic group, an indicator of occupational class. The data are drawn from a nationally representative sample of the UK population in 1987. Do those in the more prosperous classes tend to think that rates of taxation for the rich are too high, while the poor think the rates are too low? In other words, are respondents selfish about who pays for state expenditure? We can test this issue by fitting a model of no association. It does not fit (*G*-squared = 62.5 with 24 d.f., not significant), and we conclude that the variables are associated: a relationship exists between opinion and socio-economic group.

Exactly the same result would be obtained if we interchanged the first two rows of the table, or any other pair of rows, and refitted the model. (The model table counts would change correspondingly, but the *G*-squared would be unchanged.) This is because the basic no-association model assumes only categorical variables. No assumptions are made about the ordering of the categories, although it is clear from the labels of the opinion variable that in fact the categories are ordered.

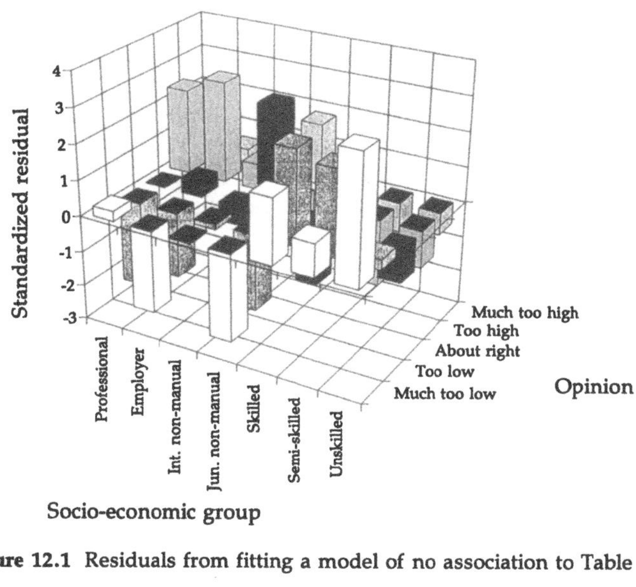

Figure 12.1 Residuals from fitting a model of no association to Table 12.1.

The residuals from fitting the no-association model (see Chapter 7) are arranged in the pattern illustrated in Figure 12.1. They are large and positive at two diagonally opposite corners of the table, and large and negative at the other two corners. These residuals indicate that the lack of fit is due to a *trend* in the data, and this can be accommodated using a model that recognizes that the variables are ordinal.

The method of creating and fitting this model is as follows. First, numeric scores are assigned to the categories of the variables. The simplest way of doing this is to number the categories in order starting from 1 upwards. For instance, we can number the opinion categories from 1 for "Much too low" to 5 for "Much too high", and the socio-economic grouping categories from 1 for "Professional" to 7 for "Unskilled".

Then a term including the product of these numeric scores is added to the basic equation for a model of no association (see Ch. 10), giving:

$$\log m_{ij} = \mu + \mu_i^A + \mu_j^B + \beta u_i v_j$$

This is the *linear-by-linear* association model. It consists of the loglinear equation for a no-association model with the addition of the term $\beta u_i v_j$. u_i represents the score assigned to the category i of the first variable, v_j is the score for category j of the second variable, and β is a constant whose value is calculated when the model is fitted. The model has one less degree of freedom than the basic no-association model because there is one more parameter to estimate, β.

Fitting this model gives a G-squared of 27.1 with 23 degrees of freedom. The fit is very good. The value of β is -0.096. One way of interpreting this figure is to notice that the logarithm of the odds ratio for any square of four adjacent cells in the table is equal to β and therefore the odds ratio is $e^{-0.096} = 0.91$. For example, the log of the ratio of the odds of semi-skilled and unskilled people thinking that taxes on high earners are far too low rather than too low is:

$$\begin{aligned}\log\left[\frac{m_{11}m_{22}}{m_{12}m_{21}}\right] &= \beta u_i v_i - \beta u_1 v_2 + \beta u_2 v_2 - \beta u_2 v_1 \\ &= \beta(u_1 - u_2)(v_1 - v_2) \\ &= \beta\end{aligned}$$

The first line comes from using the linear-by-linear model equation and cancelling out terms. The second line results from simplifying the expression on the right-hand side of the first line. Because the categories have been scored 1, 2, 3, . . . , each term in brackets is equal to 1, and the log odds ratio is therefore equal to β. If non-adjacent categories are chosen, the above expression indicates that the log odds ratio is equal to β times the differences in category scores.

If β is positive, more cases are expected to fall in cells with large scores or small scores for both variables than would be the case if there were simple independence between the variables, while if it is negative, as here, more cases would be expected in cells with small scores on one variable and large scores on the other.

Because the linear-by-linear model involves the product of the numerical scores assigned to the variables, it does not in fact, strictly speaking, assume only ordinality. The numerical scores are metric (see Ch. 2) and build into the model assumptions beyond the information available in the data. A symptom of this is that if you choose different scores, for example assigning 1, 5, 7, 9, 13 as the scores for the categories of the opinion variable, the results from fitting the model will change slightly. However, the differences in G-squared and parameter values for different scorings are not so great that they often lead to different conclusions.

The linear-by-linear model fitted the data very well. However, it required the assumption that both variables are measured on an ordinal scale. For the variable about opinion on tax rates, this seems a plausible assumption. There may be more doubt about whether the socio-economic grouping variable is ordinal. It is not very clear what underlying concept ranks the categories; it might be skill and education (but then, should junior non-manual appear in the middle, rather than at the lower end?) or it might be income (but many professionals and non-manual workers earn less than their position on the scale warrants). Whether such socio-economic and occupational class scales should be treated as ordinal or categorical has been the subject of much debate. We shall next see how we can fit a model that assumes the ordinality of the opinion variable, but that treats socio-economic grouping as categorical.

This model is known as a *column-effects* (or row-effects, depending on which is the ordinal variable) model. The equation once again has an additional term to cater for ordinality, but this time the extra term consists of the product of the numerical scores for the column ordinal variable (v_j) and a μ-term for the categorical variable (μ_i). The latter as usual consists of a set of parameters, one per row of the categorical variable, whose values are estimated when the model is fitted. The equation is:

$$\log m_{ij} = \mu + \mu_i^A + \mu_j^B + \mu_i v_j$$

The additional term in this model is the product of a μ-term and a numerical score, while the additional term in the linear-by-linear model is the product of two numerical scores.

If I is the number of categories of the row variable, A, the model has $I-1$ fewer degrees of freedom than the basic no-association model.

Fitting this model gives a G-squared of 23.4 with 20 degrees of freedom. The fit remains good. It seems that we do not need to assume that socio-economic grouping is ordinal to find a good model for these data. Whether the parameters for socio-economic group in the linear-by-linear model do in fact have the pattern expected for an ordinal variable can be tested by comparing the G-squared values for the two models, because the first model assumes ordinality for the variable while the second does not. The difference in G-squared is 2.67 with 3 degrees of freedom, indicating no significant trend arising from ordinality. We can conclude that there is no evidence from these data that socio-economic group has the trend expected if it were ordinal.

In the linear-by-linear and row- or column-effects models, we added an effect based on the scores assigned to an ordinal variable to the basic model for no association. The same can be done to augment more complex models where one or more of the variables in the table is, or is suspected to be, ordinal. A common application is where several variables are tabulated with one dimension representing time. For example, data may be obtained from successive cross-sectional surveys carried out each year on a different sample, but asking the same or similar questions. Large government surveys such as the General Household Survey and the Labour Force Survey can yield data of this form. Ordinal models can be used to investigate how the relationships between the survey variables changed over time.

Evans et al. (1991) and Payne et al. (1993) investigated the hypothesis of "class de-alignment", the idea that there has been a declining association between social class and allegiance to particular political parties, using cross-tabulations of vote by class from eight datasets each collected immediately following successive British elections. Among other models, they fitted one that includes the year of the election treated as an ordinal variable. The same kind of model may be fitted to any data table where one dimension is time and it is important to examine trends.

In Chapter 7 we examined data in which the same concept was measured at two points in time and cross-tabulated. Social mobility tables are one example of such data. In Chapter 11 we saw how causal models can be used to investigate the effects of one variable on another, where these variables are presumed to be arranged in time sequence. In this chapter we have seen how trends in relationships between concepts such as class and voting can be modelled, taking into account the fact that time is ordered. In the next chapter we will consider yet another aspect of time-related data: the modelling of durations between events.

Summary

The extra information available in ordinal variables can be used in model fitting by extending the basic loglinear model with extra terms. For two-dimensional tables, a linear-by-linear model will accommodate equally spaced ordinality in both variables, and a row- or column-effects model, ordinality in one of the variables. These models fit a trend in the data, where the counts depart from those expected if the variables were independent by increasing (or decreasing) amounts as we go along the ordinal scale. If the linear-by-linear model fits, the local odds ratio for any four adjacent cells will be constant throughout the table if the scores assigned to the variables are equally spaced one from the next.

Computer analysis

The following programs apply a linear-by-linear model and a column-effects model to the tax rate by socio-economic group data of Table 12.1. In both SPSS and GLIM, models for ordinal data can be constructed straightforwardly by adding effects that cater for the ordinal nature of one or more variables to the basic loglinear model.

SPSS

The numeric scores needed for ordinal models are usually already available because SPSS expects variables to be coded with numeric category codes. However, the LOGLINEAR command in version 4 of SPSS can handle only models in which the numeric scores are sequential and without gaps, for example, 1, 2, 3, 4, but not 1, 4, 9, 16.

In Program 12.1, after reading the data, two extra parameters are calculated, one for a linear-by-linear model and one for a column-effects model. The first (B) is the product of the numeric scores of the two variables and is used for the additional term needed for the linear-by-linear model. The second (Col) is set equal to the scores of the ordinal variable and is used for the column effects model. These additional parameters are known as *covariates* and must be included among the variables specified in the LOGLINEAR command, after the keyword WITH.

Program 12.1

```
DATA LIST FREE/ TaxHi, SEG, Count.
BEGIN DATA
1 1   5  1 2   8  1 3  13  1 4  12  1 5  33  1 6  23  1 7  12
2 1  14  2 2  73  2 3  67  2 4  99  2 5 154  2 6 119  2 7  26
3 1  28  3 2 106  3 3  78  3 4 164  3 5 145  3 6 115  3 7  26
4 1  20  4 2  62  4 3  40  4 4  58  4 5  62  4 6  43  4 7  10
5 1   5  5 2  16  5 3  12  5 4  26  5 5  16  5 6  14  5 7   3
END DATA.
WEIGHT BY Count.
* Linear by linear parameter.
COMPUTE B = TaxHi * SEG.
* Column effects parameter (Only TaxHi is assumed ordinal).
COMPUTE Col = TaxHi.
LOGLINEAR Taxhi(1,5), SEG(1,7) WITH B, Col
 /PRINT = ESTIM
 /DESIGN = TaxHi, SEG
 /DESIGN = TaxHi, SEG, B
 /DESIGN = TaxHi, SEG, Col BY SEG.
```

GLIM

The GLIM program calculates the additional parameters for the ordinal models in a way very similar to SPSS (see Program 12.2).

Program 12.2

```
$units 35
$factor TaxHi 5 SEG 7
$calculate TaxHi = %GL(5,7) : SEG = %GL(7,1)
$data Count
$read
  5   8  13  12  33  23  12
 14  73  67  99 154 119  26
 28 106  78 164 145 115  26
 20  62  40  58  62  43  10
  5  16  12  26  16  14   3
$yvar Count
$calculate B = TaxHi * SEG
$calculate Col = TaxHi
$error p
$c simple no association model
$fit TaxHi + SEG
$c linear-by-linear model
$fit TaxHi + SEG + B
```

```
$display e
$c column effects model
$fit TaxHi + SEG + Col.SEG
$display e
```

Further reading

A rather technical, but thorough, treatment of models for ordinal data can be found in Agresti (1984) and more briefly in Agresti (1990). Norusis (1990a) describes how to fit such models using SPSS. Hout (1983) discusses the application of ordinal models to social mobility data.

CHAPTER THIRTEEN
Event history models

At the end of the previous chapter, we considered the use of ordinal models for the analysis of relationships that might be changing over time. Another kind of time-related problem is where the researcher is interested in accounting for variations in the length of time between two events. For example, we might be interested in the time between marriage and divorce, between becoming unemployed and getting a job, between obtaining a job and getting promoted, and so on. The problem is to explain the interval between the events in terms of some set of explanatory variables. The time until obtaining employment may, for example, be explained by reference to educational qualifications, the state of the job market, previous employment history and other such variables.

Problems of this type can be examined using *event history analysis*. There are two basic kinds of event history analysis: those that involve discrete-time models, where the time dimension is divided up into periods such as years or months and it is assumed that changes within a time period are insignificant compared with changes between periods, and continuous-time models, where time is measured as a continuous variable and changes may occur smoothly over time. In the following, we will deal only with discrete-time models because these can be estimated using a version of logistic regression. An introduction to continuous-time analysis, which requires rather more complicated statistical models, can be found in Allison (1984). In most circumstances, the results obtained are similar regardless of whether discrete- or continuous-time assumptions are made.

The data for event history analysis is typically collected either by retrospective life histories or through prospective studies. In the former case, a sample of respondents are interviewed about aspects of their lives. For example, they may be asked about all the jobs and spells of unemployment they have experienced since leaving school. In the latter case, less usual in social research but common in medical studies, members of a sample are tracked over time and questioned every so often about what is happening to them. An example of the former, life history, approach is the Women and Employment Survey (Martin & Roberts 1984), which asked a large sample of women to provide a history of their involvement in the labour market since leaving school, including details about full-time and part-time work and child-bearing. An example of a prospective study is the British Household Panel Study at the University of Essex, which is revisiting a random sample of UK households every couple of years for a decade, starting in 1992.

The process of deciding to emigrate and then actually emigrating is another example of a topic that can be investigated using event history analysis. Massey (1987) was concerned with explaining the occurrence and timing of the migration of Mexican workers into the USA. The process of migration, he suggests, involves four segments corresponding to the fundamental decisions that migrants and their families make: whether to begin temporary visits to the USA, whether to continue visiting, whether to settle permanently in the USA, and whether eventually to return to Mexico after many years of settlement. He used a retrospective study, interviewing in 1982 a sample of 357 migrants to the USA from four Mexican communities. He asked them about their work history, their marital history, their fertility and their experience of ownership of property and businesses. For each year of a respondent's life, the occurrence of important transitions was noted and personal background characteristics recorded.

For simplicity, let us consider only the decision to settle in the USA, setting aside the other decisions involved in migration. The analytic problem is to explain the changes over time in the probability that a Mexican respondent who is living in the USA will make a decision to settle there permanently, in terms of other variables such as the migrant's household characteristics, marital status, education and so on. At the date of Massey's survey, some respondents had already taken an irrevocable decision to settle in the USA, and for these people the data will include the time when the decision was made. Others will have visited, but will not have decided to settle in the USA. For the latter group we do not know the date of the decision to settle, because the decision had not yet and may never have been made.

The data can be represented schematically as in Figure 13.1. This shows the life histories of three hypothetical respondents through time. The date of the survey is marked by the vertical line. Person A has decided to settle and the time of the decision is recorded in the survey data. Person B has not and never will settle, remaining undecided until death. Person C will decide to settle, but not until after the data collection is completed and so the decision date is not known. The fact that the dates of some events are unknown because they have not yet occurred is characteristic of event history data and is called *censoring*.

One approach to analysis might be to disregard respondents with censored data entirely. In the case of Massey's data, this would mean excluding all those who had not yet decided to settle, a substantial proportion of the total. This will lead to the loss of data from many respondents and bias in the results and so is unsatisfactory. What is needed is a method of analysis that is capable of dealing with the fact that, although we do not know when or whether people B and C in Figure 13.1 will decide to settle, we do know that they have not yet decided.

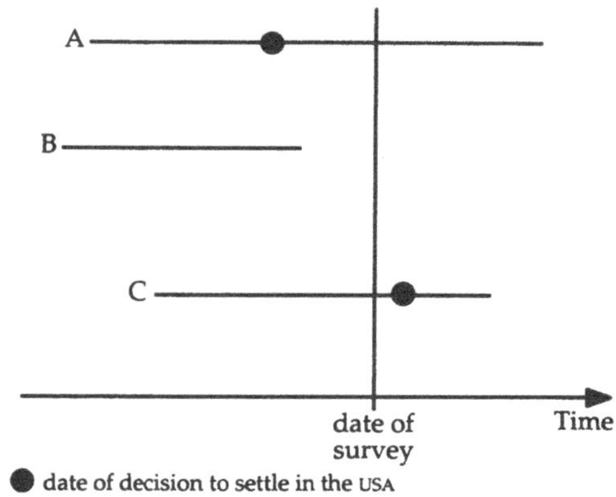

Figure 13.1 Life histories of typical migrants.

A much better approach is to recast the data so that, instead of being about the respondents, they are about the periods of time during which the respondents who are living in the USA have not yet decided to settle. In other words, the unit of analysis is changed from the respondent to the respondent-year. This involves creating a new dataset. The original dataset has one record for each respondent. The new dataset has one record for every year for each potential migrant living in the USA, until that person has made the decision to settle. For example, a migrant who moves to the USA in 1970 and in 1979 makes the decision to stay will be represented by 10 records, one for every year before the decision and one for the year of the decision. The last record is marked to indicate that the "event" – the decision to settle – happened that year. Someone who moves to the USA in 1975 and had not decided to settle by 1982, the date of the survey, will be represented by eight records, one for each year between 1975 and 1982. None of the eight records will be marked to indicate a decision to settle. The explanatory variables such as household characteristics, marital status and so on are copied on to every person-year record according to the values they took on for that person in that year.

The advantage of this form of data is that it includes records for every respondent for every year in which they were liable to make the decision to settle permanently in the USA, but no records about periods when they were not liable (e.g. because they were either not living in the USA or had already decided to settle) and no records for periods outside the scope of the survey (e.g. for times after the survey date). Thus, respondents whose date of making a decision to settle is censored contribute exactly what is known about them: that they did not make a decision before the survey.

Using these person-year data, Massey was able to discover which explanatory variables best account for the probability that a migrant living in the USA will decide to settle. He found that the probability of settlement is considerably enhanced by a lack of children and peaks among respondents in their early thirties, being lower among the younger and older migrants. The decision to settle is, therefore, most likely at the early stages of the life course, just before or after marriage, but before children have been born. Ownership of farmland, a home or a business and marital status are relatively unimportant in affecting the propensity to settle, while the number of years already lived in the USA is important: the longer the migrant has been in the USA, the higher the probability of settlement.

To illustrate in more detail how such an analysis is carried out, let us consider another example, concerned this time with the duration of spells of unemployment. In the mid-1980s, the UK Economic and Social Research Council commissioned a large programme of research into the interaction between social change and economic life, the SCEL Initiative (SCELI). The project studied changes in six urban labour markets in Britain. In each locality, about 1000 individuals aged between 20 and 60 were interviewed in 1986. Both life histories (including achievement of qualifications, marriage, changes in living arrangements, birth and departure of children from the household) and work histories (each change of job since leaving full-time education, and employment status changes, such as spells of unemployment or looking after the home) were recorded for each respondent.

Table 13.1 shows data taken from this survey, focusing on spells of unemployment. There were 6,110 respondents altogether from the six localities. These respondents were involved in a total of 3,521 separate episodes of unemployment which had either terminated in a full-time job or were still in progress at the date of the survey. All spells from the time the respondents left full-time education to the time of the survey are included, some people becoming unemployed more than once. Marsh & Gershuny (1991) discuss how the SCELI data can be manipulated using standard computer packages to yield data organised to give a record for each unemployment spell.

Each spell of unemployment was then divided into (up to) five periods: the first three months, the second three months, the next six months, the next year, and any remaining time unemployed. These periods were chosen because it may be assumed that the chance of getting a job during, for example, the second year of unemployment is much the same throughout that year, but differs from the chance of gaining employment at other times. One record was generated for each unemployed respondent-period, and the respondent's age at the time of the start of unemployment, their gender and whether or not they had any educational qualifications were attached to each record.

Table 13.1 Distribution of ceasing unemployment by time period.

Period of unemployment	Number unemployed	Number of these obtaining a full-time job	Number with censored termination	Aggregate hazard rate
1 to 3 months	3521	871	397	0.247
4 to 6 months	2253	447	183	0.198
7 to 12 months	1623	355	296	0.219
13 to 24 months	972	227	203	0.234
25 plus months	542	143	399	0.264
Total		2043	1478	0.236

Source: from SCELI data, derived by Colin Mills.

The table shows the number of unemployment spells, i.e. the number of people unemployed, included in each of the periods of unemployment. For example, of the total of 3,521 spells of unemployment, 871 ceased during the first three months because the respondent took on a full-time job, and 397 were censored; that is, the three-month period was still continuing at the time of the survey, so the fate of the respondent is not known. The remaining 2,253 spells continued into the second period (4-6 months). Of these, 447 terminated with full-time employment during that period; 183 had a censored end point and 1,623 continued into the third period. And so on. In all, 2,043 of the 3,521 spells ended with the unemployed person taking up a full-time job and 1,478 were still on-going at the time of the survey and have therefore to be treated as censored.

As each individual moves through a spell of unemployment, there is a probability, changing over time, that that person will get a job. This probability is known as the *hazard rate*. The term may seem inappropriate for the chance of getting a job, but it comes from medical statistics where the event being considered is often death from disease. The hazard rate is defined as the chance that an event will occur during a particular period to a particular individual, given that that person is at risk of the hazard at that time.

The hazard rate is a characteristic of an individual at a given time. The changing hazard rate aggregated for the whole sample is shown in the last column of Table 13.1. These values are obtained by dividing the number ending with a full-time job by the total number in any period. For example, during the first three months, there was a chance of 871 in 3,521, or about 25 per cent, that the spell would end with a job. After the first three months, the hazard of getting a job falls sharply and then slowly rises. Unemployed people have most chance of getting a job during the first three months: about 25 per cent do. The likelihood of getting a job after four to six months of unemployment is smaller, around 20 per cent. After that, the chance of employment increases slowly in each successive period.

These conclusions about changes in the hazard rate have been obtained from considering the overall distribution of the hazard over periods for the whole sample. We now turn to examining how the hazard rate for individuals varies with that individual's personal circumstances. For example, it seems likely that the hazard of employment will differ between those with and without educational qualifications. Those with qualifications may be more likely to obtain a job within the first three months. Other important explanatory variables could be the respondent's age and gender. We can set up a model with the hazard rate as the dependent variable and age, gender and qualifications as explanatory variables. In addition, we shall include the duration of unemployment because, as we have seen, the hazard rate seems to depend on how long the person has been unemployed.

This model will be applied to a five-dimensional table cross-tabulating period of unemployment, whether the respondent got a job in that period and the respondent's age, gender and qualifications. Table 13.1 was a marginal table from this data table, obtained by summing over the age, gender and qualification variables. As with Table 13.1, the counts in the data table are not of respondents, but of respondent-periods. The data table includes five variables and 120 cells and is too large to display (but see Program 13.2).

To analyze the data, logistic regression can be used as in Chapter 10, with the hazard rate as the dependent variable and the others as the independent variables. A main-effects model is close to giving a good fit, with a G-squared of 79.4 and 51 degrees of freedom. However, to bring the fit up to the 5 per cent criterion, two interactions have to be added: an interaction relating the hazard rate, the period and the respondent's gender, and an interaction between the hazard rate, the period and the respondent's qualifications (G-squared = 57.0, d.f. 45). There is a slight tendency for the change in the hazard of gaining a full-time job as the duration of unemployment lengthens to be different for men and women and for those with and without qualifications.

If we just look at the parameters for the main effects, expressed as odds ratios (Table 13.2), we can see that, relative to the hazard of getting a job during the first three months of unemployment, the chance of employment drops in the second three months and then steadily climbs again. It seems that, if you become unemployed and cannot step very quickly into another job, your chances of finding a job fall very sharply and gradually increase thereafter. There is no sign in these data of any significant extra disadvantage resulting from becoming one of the "long-term" unemployed after six months or a year.

Age is, however, a disadvantage in getting a job. Those aged over 30 have only two-thirds the probability of getting a job compared with those aged 20 and under. Being female is also a disadvantage, while possessing some qualifications provides a significant advantage, increasing the hazard rate by 50 per cent.

The interaction terms required for a model that fits the data identify four circumstances in which the hazard rate is significantly different from that predicted by the main effects only. Those who have qualifications have an increased chance of employment (by a factor of 1.3) in the second period, i.e. between four and six months after becoming unemployed, and a decreased chance (by 0.7) in the fourth period, i.e. during the second year of unemployment, compared with those without qualifications. Women have a lower probability of becoming employed after they have reached the age of 20: about three-quarters the rate they would be expected to experience, taking account only of the separate effects of age and gender.

Table 13.2 Parameter estimates for main effects model of hazard of obtaining full-time employment.

	Odds ratio
Period of unemployment	
1 to 3 months	•
4 to 6 months	0.69
7 to 12 months	0.90
13 to 24 months	*1.12*
25 plus months	1.26
Age	
Up to 20	•
21 to 30	*0.97*
Over 30	0.65
Gender	
Male	•
Female	0.81
Qualifications	
No qualifications	•
Some qualifications	1.47

Key: • reference category
Note: Values in *italics* are not significantly different from 1 at the 5% level.

The major source of difficulty in doing this kind of analysis is that data rarely comes already prepared as person-period records. More usually, data are available with one record for each respondent and the researcher has then to manipulate them to the person-period records required for event history analysis. This can mean writing special computer programs or using sophisticated data management packages, skills not usually associated with social research. If this problem can be overcome, however, event history analysis can be a very powerful tool for examining change.

The examples we have discussed in this chapter have avoided some of the complications that can be encountered in event history analysis.

For instance, some respondents in the SCELI data were unemployed more than once, yet we have treated the second and subsequent spells as being no different from the first. Although this is not likely to be true (because the experience gained when first unemployed will probably affect later spells and the chances of re-employment), it made the analysis much more straightforward. To take account properly of the fact that an event may be *repeated* requires more complicated models. A relatively simple way of dealing with repeated events is to add an additional explanatory variable to the model: the number of events that the respondent has experienced before the current spell.

A second complication occurs when there are *competing* events. For instance, we may want to model the fact that a spell of unemployment may end with the person either finding a job or withdrawing from the labour market. The best way of tackling this is to treat each type of event separately. Thus, we first model the hazard of getting a job, treating those spells that terminate with the respondent withdrawing from the labour market as though they were censored. Then we do the same for spells where the respondent has left the labour market, treating spells that end with getting a job as censored. The two models may turn out to include different effects and have different parameter estimates, reflecting different underlying processes leading to the two types of event.

A third difficulty is *left censoring*. This occurs when the time of the beginning of a spell is not known. (Compare the *right censoring* described above, which occurs when the time of the end of a spell is not known.) For example, a prospective study may follow people through their lives, recording spells of unemployment. However, some people will be unemployed at the time the study begins and it may not be possible to establish accurately when their unemployment began. In practice, the best way of dealing with left censoring is to discard left-censored person-period records from the analysis. While this wastes some data, it is difficult to do anything useful with such records.

Summary

Event history analysis studies the likelihood that an event, such as employment or unemployment, marriage, the birth of a child or migration, will occur to a person, and relates this "hazard rate" to explanatory variables. If the hazard rate can be assumed to be constant during discrete periods of time, logistic regression can be applied to estimate the parameters of an event history model. This requires the data to be transformed into one record for each period for each respondent. A table of whether the event occurred during the period cross-tabulated by the explanatory variables is the basis for the logistic regression.

Computer analysis

Programs 13.1 and 13.2 use logistic regression to fit a model and estimate the parameters for the SCELI unemployment data discussed above.

SPSS

Program 13.1 reads the data, labels the variables and displays the cross-tabulation of Period by Event shown in Table 13.1. A logistic regression then fits a model to the data and displays the parameter estimates.

Program 13.1

```
TITLE Program 13.1
DATA LIST FREE / Count, Event, Period, Age, Gender, Qual.
BEGIN DATA.
279 1 1 1 1 1
119 2 1 1 1 1
194 1 2 1 1 1
 56 2 2 1 1 1
[Lines of data omitted ]
  4 2 3 3 2 2
 24 1 4 3 2 2
  2 2 4 3 2 2
 10 1 5 3 2 2
  4 2 5 3 2 2
END DATA.
WEIGHT BY Count.
VARIABLE LABELS Event 'Periods with a job'/
    Qual 'Educational qualifications'/.
VALUE LABELS
   Event  1 'Unemployment continues or is censored'
          2 'Unemployment ends with job'/
   Period 1 '1 to 3 months'  2 '4 to 6 months'
          3 '7 to 12 months' 4 '13 to 24 months'
          5 '25 months and over'/
   Age    1 'Up to 20 years' 2 '21 to 30 years'
          3 '31 years and older'/
   Gender 1 'Male'           2 'Female'/
```

```
   Qual    1 'No qualifications'
           2 'Some qualifications'/.
CROSSTABS Period BY Event
 /CELLS= COUNT, COL.
LOGLINEAR Event(1,2) BY Period(1,5), Age(1,3),  Gender(1,2),
                         Qual(1,2)
 /PRINT = ESTIM
 /DESIGN = Event, Event BY Period, Event BY Age,
           Event BY Gender,
           Event BY Qual, Event BY Period BY Qual,
           Event BY Age BY Gender.
```

GLIM

Program 13.2 fits the logistic model using the equivalent loglinear model. The estimates are those shown in Table 13.2.

Program 13.2

```
$units 120
$factor Event 2 Period 5 Age 3 Gender 2 Qual 2
$calc Event = %GL(2,1) : Period = %GL(5,2) : Age = %GL(3,10)
$calc Gender = %GL(2, 30) : Qual = %GL(2, 60)
$data Count
$read
279  119  194   56  145   35   85   33   48   23  327  119
245   61  154   65   96   32   50   29  403   87  305   47
225   48  143   45   80   27  300  102  227   42  163   48
 95   37   58   16  215   42  162   21  129   19   76   17
 44   12  192   15  142   20  112   14   71    9   39    4
153   80   92   30   54   28   24   14    9    5  202  100
105   53   72   25   41   13   15   10  110   40   61   24
 36   15   19    4   12    0  155   64   96   32   65   21
 36   13   18    9  227   83  129   51   79   33   35    8
 16    4   87   20   48   10   34    4   24    2   10    4
$yvar Count
$error P
$fit Event*Period*Qual + Event*Age*Gender +
     Period*Age*Gender*Qual
$display e
```

Further reading

A book and a chapter by Allison (1984, 1982) provide a readable introduction to event history analysis, the latter emphasizing discrete-time models. More wide-ranging, accessible introductions can be found in Blossfield et al. (1989) and Yamaguchi (1991). Karweit & Kertzer (1986) discuss creating and managing event history data records. Briefer reviews of current developments can be found in Peterson (1991) and Hutchinson (1988a,b).

REFERENCES

Abell, P. 1971. *Model building in sociology*. London: Weidenfeld & Nicolson.

Agresti, A. 1984. *Analysis of ordinal categorical data*. New York: John Wiley.

Agresti, A. 1990. *Categorical data analysis*. New York: John Wiley.

Aitkin, M., D. Anderson, B. Francis, J. Hinde 1989. *Statistical modelling in GLIM*. Oxford: Oxford University Press.

Alba, R. D. 1988. Interpreting the parameters of log-linear models. In *Common problems/proper solutions: avoiding errors in quantitative research*, J. S. Long (ed.), 258–87. Beverly Hills, CA: Sage.

Allison, P. D. 1982. Discrete-time methods for the analysis of event histories. In *Sociological methodology 1982*, S. Leinhardt (ed.). San Francisco: Jossey-Bass.

Allison, P. D. 1984. *Event history analysis*. Beverly Hills, CA: Sage.

Arber, S. & G. N. Gilbert 1989. Men: the forgotten carers. *Sociology* **23**, 111–18.

Arber, S. L. & L. Sawyer 1979. Changes in the structure of general practice: the patient's viewpoint. Unpublished report, Department of Health and Social Security.

Atkins, L. & D. Jarrett 1979. The significance of "significance tests". In *Demystifying social statistics*, J. Irvine, A. Miles, J. Evans (eds). London: Pluto.

Babbie, E. R. 1973. *Survey research methods*. Belmont, CA: Wadsworth.

Benedetti, J. K. & M. B. Brown 1978. Strategies for the selection of log-linear models. *Biometrics* **34**, 680–86.

Bishop, Y. M. M., S. E. Fienberg, P. W. Holland 1975. *Discrete multivariate analysis*. Cambridge, MA: MIT Press.

Blalock, H. M. 1964. *Causal inferences in non-experimental research*. Chapel Hill, NC: University of North Carolina Press.

Blalock, H. M. 1979. *Social statistics*. New York: McGraw-Hill.

Blalock, H. M. 1982. *Conceptualization and measurement in the social sciences*. Beverly Hills, CA: Sage.

Blau, P. M. & O. D. Duncan 1967. *The American occupational structure*. New York: John Wiley.

Blossfield, H. P., A. Hamerle, K. U. Mayer 1989. *Event history analysis*. Hillsdale, NJ: Lawrence Erlbaum.

Breeze, E., G. Trevor, A. Wilmot 1991. *General Household Survey, 1989*. London: HMSO.

Bryman, A. & D. Cramer 1990. *Quantitative data analysis for social scientists*. London: Routledge.

Cartwright, A. 1976. *How many children?* London: Routledge & Kegan Paul.

Caulcott, E. 1973. *Significance tests*. London: Routledge & Kegan Paul.

Clogg, C. & S. Eliason 1988. Some common problems in log-linear models. In *Common problems/proper solutions: avoiding errors in quantitative research*, J. S.

Long (ed.). Beverly Hills, CA: Sage.
Cowart, A. T. 1973. Electoral choice in American states: incumbency effects, partisan forces and divergent partisan majorities. *American Political Science Review* **67**, 835–53.
Crosby, C. 1978. *Intra-urban migration*. York: University of York Press.

Dale, A., S. Arber, M. Procter 1988. *Doing secondary analysis*. London: Unwin Hyman.
Davis, J. A. 1973. Hierarchical models for significance tests in multivariate contingency tables. In *Sociological Methodology*, H. L. Costner (ed.), 189–231. San Francisco: Jossey-Bass.
Duncan, O. D. 1965. Partitioning polytomous variables in multiway contingency analysis. *Social Science Research* **4**, 167–82.
Duncan, O. D. 1975. *Introduction to structural equation models*. New York: Academic Press.

Erikson, R. & J. H. Goldthorpe 1978. Commonality and variation in social fluidity in industrial nations. Part I, A model for evaluating the FJH hypothesis. *European Sociological Review* **3**, 54–77.
Erikson, R. & J. H. Goldthorpe 1992. *The constant flux*. Oxford: Oxford University Press.
Evans, G., A. Heath, C. Payne 1991. Modelling trends in the class/party relationship 1964–87. *Electoral Studies* **10**, 99–117.

Fienberg, S. E. 1980. *The analysis of cross-classified categorical data*, 2nd edn. Cambridge, MA: MIT Press.
Francis, B., M. Green, C. Payne (eds) 1993. *The GLIM System Release 4 Manual*. Oxford: Oxford University Press.
Frude, N. 1987. *A guide to SPSS/PC+*. London: Macmillan.

Gilbert, G. N. 1981. *Modelling society*. London: Allen & Unwin.
Gilbert, G. N. (ed.) 1992. *Researching social life*. London: Sage.
Goldthorpe, J. H. 1980. *Social mobility and class structure in modern Britain*. Oxford: Oxford University Press.
Goodman, L. A. 1965. On the statistical analysis of mobility data. *American Journal of Sociology* **71**, 290–301.
Goodman, L. A. 1972. A modified multiple regression approach to the analysis of dichotomous variables. *American Sociological Review* **37**, 28–46.
Goodman, L. A. 1973a. The analysis of multidimensional contingency tables when some variables are posterior to others: a modified path analysis approach. *Biometrika* **60**, 179–92.
Goodman, L. A. 1973b. Causal analysis of data from panel studies and other kinds of surveys. *American Journal of Sociology* **78**, 1135–91.
Goodman, L. A. & W. H. Kruskal 1954. Measures of association for cross-classifications. *Journal of the American Statistical Association* **49**, 732–64.

Haberman, S. J. 1978. *The analysis of qualitative data*. New York: Academic Press.
Haskey, J. 1992. Pre-marital cohabitation and the probability of subsequent

divorce: analyses using new data from the General Household Survey. *Population Trends* **68**, 10–19.
Hauser, R. M. 1978. A structural model of the mobility table. *Social Forces* **56**, 919–53.
Hauser, R. M. 1979. Some exploratory methods for modelling mobility tables and other cross-classified data. In *Sociological Methodology, 1980*, K. F. Schuessler (ed.). San Francisco: Jossey-Bass.
Healy, M. J. R. 1988. *GLIM: an introduction*. Oxford: Oxford University Press.
Heise, D. R. 1975. *Causal analysis*. New York: John Wiley.
Holt, D., T. M. F. Smith, P. D. Winter 1980. Regression analysis of data from complex surveys. *Journal of the Royal Statistical Society* **A143**, 474–87.
Hosmer, D. W. & S. Lemeshow 1989. *Applied logistic regression*. New York: John Wiley.
Hout, M. 1983. *Mobility tables*. Quantitative applications in the social sciences. Beverly Hills, CA: Sage.
Hutchinson, D. 1988a. Event history and survival analysis in the social sciences. I, Background and introduction. *Quality and Quantity* **22**, 203–19.
Hutchinson, D. 1988b. Event history and survival analysis in the social sciences. II, Advanced applications and recent developments. *Quality and Quantity* **22**, 255–78.

ICPSR 1977. *National data program for the social sciences General Social Survey*. Inter-Universities Consortium for Political and Social Research, Institute for Social Research of the University of Michigan.
ICPSR 1978. *National data program for the social sciences General Social Survey*. Inter-Universities Consortium for Political and Social Research, Institute for Social Research of the University of Michigan.

Jowell, R., S. Witherspoon, L. Brook 1988. *British social attitudes: the 5th report*. Aldershot, England: Gower Press.

Karweit, N. & D. I. Kertzer 1986. Database management for life course family research. *Current Perspectives of Ageing and the Life Cycle* **2**, 167–88.
Knoke, D. & P. J. Burke 1980. *Log-linear models*. Beverly Hills, CA: Sage.

Lipset, S. M., M. Trow, J. Coleman 1956. *Union democracy*. New York: Free Press.
Loether, H. J. & D. G. McTavish 1992. *Descriptive and inferential statistics: an introduction*, 4th edn. Boston: Allyn & Bacon.

Mack, J. 1978. Race and the census. *New Society*, 27 July, 191.
Marsh, C. & J. Gershuny 1991. Handling work history data in standard statistical packages. In *Life and work history analyses: qualitative and quantitative developments*, S. Dex (ed.). London: Routledge.
Martin, J. & C. Roberts 1984. *Women and employment: a lifetime perspective*. London: HMSO.
Massey, D. S. 1987. Understanding Mexican migration to the United States. *American Journal of Sociology* **92**, 1372–403.

Merton, R. K. 1968. Contributions to the theory of reference group behaviour. In *Social theory and social structure*, R. K. Merton (ed.). Glencoe, IL: Free Press.

Norusis, M. J. 1990a. *SPSS/PC+ 4.0 advanced statistics guide*. Chicago: SPSS.
Norusis, M. J. 1990b. *SPSS/PC+ 4.0 base manual*. Chicago: SPSS.

Office of Population Censuses and Surveys (OPCS) 1970. *Classification of occupations*. London: HMSO.
Office of Population Censuses and Surveys (OPCS) 1978. *OPCS Monitor*. Vol. CEN 78/4. London: OPCS.
Office of Population Censuses and Surveys 1990. *General Household Survey, 1980*. Colchester: ESRC Data Archive.
Oppenheim, A. N. 1991. *Questionnaire design and attitude measurement*. London: Heinemann.

Payne, C. D. (ed.) 1986. *The GLIM system release 3.77 manual*. Oxford: Numerical Algorithms Group.
Payne, C., J. Payne, A. Heath 1993. Modelling trends in multiway tables. In *Analysing social and political change: a casebook of methods*, R. Davies & A. Dale (eds). London: Sage.
Peterson, T. 1991. The statistical analysis of event histories. *Sociological methods and research* **19**, 270–323.

Reynolds, H. T. 1977. *The analysis of cross-classifications*. New York: Free Press.
Rosenberg, M. 1968. *The logic of survey analysis*. New York: Basic Books.
Runciman, W. G. 1966. *Relative deprivation and social justice*. London: Routledge & Kegan Paul.

SCPR 1987. *British social attitudes survey*. London: Social and Community Planning Research.
Skinner, C. J., D. Holt, T. M. F. Smith (eds) 1989. *Analysis of complex surveys*. Chichester: John Wiley.
Stacey, M., E. Batstone, C. Bell, A. Murcott 1975. *Power, persistence and change*. London: Routledge & Kegan Paul.
Stier, H. & D. B. Grusky 1990. An overlapping persistence model of career mobility. *American Sociological Review* **55**, 736–56.
Stouffer, S. A., E. A. Suchman, L. C. Devinney, S. A. Star, R. M. J. Williams 1949. *The American soldier*, Vol. 1. Princeton, NJ: Princeton University Press.

Thornes, B. & J. Collard 1979. *Who divorces?* London: Routledge & Kegan Paul.

Willer, D. E. 1967. *Scientific sociology*. Englewood Cliffs, NJ: Prentice-Hall.
Wrigley, N. 1985. *Categorical data analysis for geographers and environmental scientists*. London: Longman.

Yamaguchi, K. 1991. *Event history analysis*. Beverly Hills, CA: Sage.

Zeisel, H. 1958. *Say it with figures*. London: Routledge & Kegan Paul.

INDEX

For Product Safety Concerns and Information please contact our
EU representative GPSR@taylorandfrancis.com Taylor & Francis
Verlag GmbH, Kaufingerstraße 24, 80331 München, Germany